SOIL PH FOR NUTRIENT AVAILABILITY AND CROP PERFORMANCE

Edited by **Suarau Oshunsanya**

Soil pH for Nutrient Availability and Crop Performance

http://dx.doi.org/10.5772/68057

Edited by Suarau Oshunsanya

Contributors

Danute Karcauskiene, Regina Repsiene, Dalia Ambrazaitiene, Regina Skuodiene, Ieva Jokubauskaite, Enos Wambu, Shinji Matsumoto, Hideki Shimada, Takashi Sasaoka, Ginting Kusuma, Rudy Gautama, Ikuo Miyajima, Suarau Oshunsanya

Notice

Statements and opinions expressed in the chapters are these of the individual contributors and not necessarily those of the editors or publisher. No responsibility is accepted for the accuracy of information contained in the published chapters. The publisher assumes no responsibility for any damage or injury to persons or property arising out of the use of any materials, instructions, methods or ideas contained in the book.

First published in London, United Kingdom, 2019 by IntechOpen

IntechOpen is the global imprint of INTECHOPEN LIMITED, registered in England and Wales, registration number: 11086078, The Shard, 25th floor, 32 London Bridge Street
London, SE19SG – United Kingdom
Printed in Croatia

British Library Cataloguing-in-Publication Data
A catalogue record for this book is available from the British Library

Additional hard copies can be obtained from orders@intechopen.com

Soil pH for Nutrient Availability and Crop Performance, Edited by Suarau Oshunsanya
p. cm.
Print ISBN 978-1-78985-015-4
Online ISBN 978-1-78985-016-1

We are IntechOpen,
the world's leading publisher of
Open Access books
Built by scientists, for scientists

4,000+
Open access books available

116,000+
International authors and editors

120M+
Downloads

Our authors are among the

151
Countries delivered to

Top 1%
most cited scientists

12.2%
Contributors from top 500 universities

Interested in publishing with us?
Contact book.department@intechopen.com

Numbers displayed above are based on latest data collected.
For more information visit www.intechopen.com

Meet the editor

S.O. Oshunsanya is a senior lecturer in the Department of Agronomy, University of Ibadan, Nigeria. His research focus is primarily on soil and water conservation and soil amendment for sustainable crop cultivation. He has taught soil science related courses at both graduate and postgraduate levels for 10 years, and has been involved in many research and community services as a consultant in Nigeria and China. He has supervised over 20 postgraduate students and published over 60 articles. He has served the University of Ibadan at various capacities both within and outside and is a member of learned societies at both national and international levels.

Contents

Preface

Soil pH is a measure of a soil solution's acidity and alkalinity, which affects nutrient solubility and availability in the soil. Soil pH is affected by the use of nitrogen fertilizers, organic matter decomposition, soil types and land use, rainfall, weathering of minerals, and parent material. Soil pH affects soil productivity by regulating the availability of nutrients for crop plant uptake. Soil pH levels near 7 are optimal for overall nutrient availability, crop tolerance, and soil microorganism activity. Crop productivity starts to decrease when the soil pH falls below 5.5 ($CaCl_2$). Toxic amounts of aluminum and manganese are soluble and released into the soil solution when soil pH falls below 5.0 ($CaCl_2$). Soluble aluminum is toxic to roots of many crop plants and therefore limits their access to soil, water, and nutrients. At a low soil pH, microbial activity decreases and nutrients such as phosphorus, magnesium, calcium, and molybdenum gradually become unavailable to the roots of crops in the soil. At this stage, fertilizers become less effective and agricultural production can be significantly reduced. As the soil becomes acidic, fewer agricultural crops grow. Soil pH can be modified by using soil amendments. Lime or lime compounds (ground limestone, marl, and hydrated lime) are the soil amendments used to raise the soil pH or increase the soil acidity. Sulfur is used to lower soil pH or increase the acidity of the soil. In areas of high total rainfall, calcium and magnesium could be washed from the soil resulting in an acid soil. Lime can also be used to replace the lost calcium and raise the soil pH to a range preferred by the crop being cultivated.

The editor wishes to place on record the unquantifiable assistance rendered by Ms. Kristina Kardum.

S.O. Oshunsanya
B. Agric, M. Sc., PhD
Department of Agronomy
Faculty of Agriculture
University of Ibadan, Nigeria

Introduction

Chapter 1

Introductory Chapter: Relevance of Soil pH to Agriculture

Suarau Odutola Oshunsanya

Additional information is available at the end of the chapter

http://dx.doi.org/10.5772/intechopen.82551

1. Why soil pH?

Soil pH is a master variable in soils because it controls many chemical and biochemical processes operating within the soil. It is a measure of the acidity or alkalinity of a soil. The study of soil pH is very important in agriculture due to the fact that soil pH regulates plant nutrient availability by controlling the chemical forms of the different nutrients and also influences their chemical reactions. As a result, soil and crop productivities are linked to soil pH value. Though soil pH generally ranges from 1 to 14, the optimum range for most agricultural crops is between 5.5 and 7.5. However, some crops have adapted to thrive at soil pH values outside this optimum range. The United States Department of Agricultural National Resources Conservation Service groups soil pH values as follows: ultra acidic (<3.5), extremely acidic (3.5–4.4), very strongly acid (4.5–5.0), strongly acidic (5.1–5.5), moderately acidic (5.6–6.0), slightly acidic (6.1–6.5), neutral (6.6–7.3), slightly alkaline (7.4–7.8), moderately alkaline (7.9–8.4), strongly alkaline (8.5–9.0) and very strongly alkaline (>9.0) [1].

Soil pH is affected by the mineral composition of the soil parent material and the weathering reactions undergone by that parent material. For instance, in humid environments, soil acidification occurs for a long time as the products of weathering leached by water moving laterally or downwards through the soil, while in the dry environments, soil weathering and leaching are less intense, and soil pH is often neutral or alkaline [2].

2. Soil acidification

Soil acidification is brought about by a number of processes such as high rainfall, crop growth, the use of fertilizers, acid rain and oxidative weathering.

2.1. High rainfall

Soils usually become acidic under heavy rainfall. This is because rainwater is slightly acidic (about 5.7) due to a reaction with CO_2 in the atmosphere that forms carbonic acid. As this rainwater passes through soil pores, it leaches basic cations from the soil as bicarbonates, which increases the percentage of Al^{3+} and H^+ relative to other cations in the soil. Root respiration and decomposition of organic matter by microorganisms also release CO_2 that increases the carbonic acid (H_2CO_3) concentration resulting to leaching.

2.2. Crop growth

Soil nutrients are taken up by crop roots in the form of ions (NO_3^-, NH_4^+, Ca^{2+} and $H_2PO_4^-$). Crop roots often take up more cations than anions. But crops must maintain a neutral charge in their roots for normal physiological processes to take place. H^+ ions are released by root crops to compensate for the extra positive charges resulting to acid soils.

2.3. Use of fertilizers

Some fertilizers such as ammonium (NH_4^+) fertilizers undergo nitrification process to form nitrate (NO_3^-), and during this process, H^+ ions are released leading to acid soils.

2.4. Acid rain

Oxides of sulfur and nitrogen are released into the atmosphere when burning fossil fuels. Released oxides react with rainwater in the atmosphere to form tetraoxosulphate (vi) acid and trioxonitrate (v) acid.

2.5. Oxidative weathering

Sulphides and other compounds containing Fe^{2+} produced acidity during oxidation process.

3. Soil alkalinity

Soil alkalinity increases with weathering of silicate, aluminosilicate and carbonate mineral compounds that contain Na^+, Ca^+, Mg^{2+} and K^+. The fore-listed minerals are usually added to the soil by the deposition of eroded sediments by water or wind. Soil alkalinity can also be increased by addition of water containing dissolved bicarbonates especially when irrigating with high-bicarbonate water. Insufficient water flowing to leach soluble salts can lead to accumulation of alkalinity in a soil. This is common in arid areas or poor internal soil water drainage situations, where the water that comes in is either transpired by crops or evaporates rather than flowing through the soil.

Both acid and alkaline soils have influence on crop growth and development. For instance, agricultural crops grown in acid soils may experience some stresses such as AI, H and Mn toxicity as well as Ca and Mg nutrient deficiencies. Aluminum toxicity, which is the most widely spread problem of acid soils, occurs when aluminum is present in ionic Al^{3+} form. Aluminum ion Al^{3+} is the most soluble of all forms of aluminum at soil pH less than 5.0 (acidic condition). Aluminum is not a plant nutrient but an ionic form of Al^{3+} that enters crop roots passively through the process of osmosis. Aluminum inhibits root growth and development by interfering uptake and transport of essential nutrients, cell division, cell wall formation and enzyme activity. However, strong alkaline soils (sodic soils) are characterized with slow infiltration, reduced hydraulic conductivity and poor soil water retention capacity that make crops to experience water stress.

Generally, agricultural crops are varied in terms of suitability for soil pH range. Some crops can be intolerant of a particular soil pH due to a particular mechanism. For instance, soil pH 5.5 is not suitable for soybean plants when molybdenum is low in the soil, but the same pH 5.5 becomes optimum for soybean when molybdenum is sufficient in the soil. Most agricultural crops perform optimally around soil pH 7.0 (neutral). This shows that it is very important to bring both acidic and alkaline soils to neutral soil pH value for optimum performance of crops.

4. Amendment of soil acidity and alkalinity

The pH of acidic soil can be increased by using finely ground agricultural lime (limestone or chalk). The buffering capacity of the soil determines the amount of lime needed to increase pH of acidic soil. The buffering capacity of the soil largely depends on the amount of clay and organic matter present. Soils with high clay and organic matter will have high buffering capacity. Apart from limestone, other amendments such as wood ash, industrial calcium oxide (burnt lime), magnesium oxide, basic slag (calcium silicate) and oyster shells can be used to increase pH of acidic soils. On the other hand, the pH of alkaline soils can be decreased by using acidifying fertilizers or organic materials. Acidifying fertilizers include ammonium sulphate, ammonium nitrate and urea, while acidifying organic materials are peat or sphagnum peat moss. Elemental sulfur (90–99% S) has been successfully used at application rates of 300–500 kg ha^{-1} to reduce the pH of an alkaline soil. Therefore, farmers must be encouraged to regulate the soil pH value for optimal crop performance.

Author details

Suarau Odutola Oshunsanya[1,2]

Address all correspondence to: soshunsanya@yahoo.com

1 Department of Agronomy, University of Ibadan, Nigeria

2 Key Laboratory of Agro-Environment and Agro-Product Safety, Guangxi University, Nanning, China

References

[1] Soil Survey Staff. Soil survey laboratory methods manual. In: Burt R, editor. Soil Survey Investigations Report No. 42, Version 5.0. 5th ed. U.S. Department of Agriculture, Natural Resources Conservation Service. 2014. pp. 276-279

[2] Bloom PR, Skyllberg U. Soil pH and pH buffering. In: Huang PM, Li Y, Sumner ME, editors. Handbook of Soil Sciences: Properties and Processes. 2nd ed. Boca Raton, FL: CRC Press; 2012. pp. 19-14. ISBN: 9781439803059

Soil Fertility and Plant Nutrition

Effects of Acid Soils on Plant Growth and Successful Revegetation in the Case of Mine Site

Shinji Matsumoto, Hideki Shimada,
Takashi Sasaoka, Ikuo Miyajima,
Ginting J. Kusuma and Rudy S. Gautama

Additional information is available at the end of the chapter

http://dx.doi.org/10.5772/intechopen.70928

Abstract

Acid soils are caused by mining, potentially causing the death of plants. Although soil pH is one of the useful indicators to evaluate acid soil conditions for successful revegetation, the dissolution of harmful elements under acidic conditions should be considered in addition to the tolerance mechanism of plants in mines. Thus, this study aims to report the current situation of acid soils and plant growth in mine site and to elucidate the effects of acid soils on plant growth over time through field investigation and a vegetation test. The results showed that the dissolution of Al from acid soils which were attributed to the dissolution of sulfides influenced plant growth. Not only soil pH but also the assessment of the dissolution of sulfides over time is crucial for successful revegetation, suggesting that net acid producing potential (NAPP) and net acid generation (NAG) pH, which are used for evaluating the formation of acidic water, are useful to evaluate soil conditions for the revegetation. Furthermore, acid-tolerant plant survived under acidic conditions by increasing the resistance against acidic conditions with the plant growth. Such factors and the proper selection of plant species play an important role in achieving successful revegetation in mines.

Keywords: acid soils, plant growth, revegetation, mine site, acidic water, Al tolerance, sulfides

1. Introduction

Acid soils are formed with human activities, such as construction and mining, potentially causing the death of plants [1]. Plants wither due to not only low pH conditions in acid soils

IntechOpen

but also dissolution of harmful elements, such as Al, Fe, and Mn, dissolving under acidic conditions. Although soil pH is one of the useful indicators to evaluate acid soil conditions for successful revegetation and/or farming, the dissolution of harmful elements under acidic conditions should be taken into account [2]. For example, Al, which constitutes approximately 7% of the Earth's mass, is easily released in water with the change of pH, thereby inhibiting plant growth, including root growth and its function [3]. Al generally existing as $Al(OH)_3$, which is insoluble in soils, dissolves in water as Al^{3+} under acidic conditions (pH < 4.5) and is released as $Al(OH)_4^-$ under alkaline conditions. The Al^{3+} easily reacts with phosphoric acid and then it causes phosphorus deficiency on plants with the formation of insoluble aluminum phosphate in soils [4]. Other harmful elements, such as Fe and Mn, also inhibit plant growth. Therefore, acid soils affect plant growth through indirect factors like dissolution of harmful elements, indicating that the understanding of the effect of acid soils on plant growth in terms of not only soil pH but also harmful elements is important for successful revegetation.

With respect to plant species, there are some plants that can survive in acid soils owing to tolerance characteristics, such as acid tolerance and Al tolerance. Acid-tolerant plants can survive under low soil pH conditions by setting up several tolerance mechanisms, such as the increase of soil pH around the root apices [5, 6]. The plants with Al tolerance are resistant to the effects of Al as described above. They are separated into Al-tolerant plant and Al-stimulated plant and additionally categorized as Al-excluders, Al root-accumulators, and Al-accumulators [7]. While *Camellia sinensis* localizes Al in the cell walls of epidermal cells of the leaves against Al toxicity [8], *Acacia mangium (A. mangium)* excludes Al from the root apices [9]. Furthermore, Saifuddin et al. found that the photosynthetic rates rose by increasing soil pH in *Leucaena leucocephala* after the pre-aluminum treatment [9, 10]. Thus, Al-tolerant mechanism of plants depends on the plant species, and the effects of acid soils on plant growth change according to the species. This indicates that tolerance characteristics of plants should be considered in regard to the effect of acid soils on plant growth in addition to soil pH.

Figure 1. Death of plants under acidic condition in the waste dump in mine site.

The research on the effects of acid soils on plant growth has been performed in the world; however, it is still a serious problem, especially in mine site where revegetation is necessary for environmental reclamation as shown in **Figure 1**. The formation of acid soils resulted from construction and/or resource exploitation has been often highlighted as the cause of plant death. The information on the effects of acid soils on plant growth in mine site is crucial for successful revegetation. Therefore, this study aims to report the current situation of acid soils and plant growth in mine site and to elucidate the effects of acid soils on plant growth over time through field investigation and a vegetation test. On the basis of the results, the key to successful revegetation was discussed from the point of view of soil acidification and tolerance mechanism of plants.

2. Methods

2.1. Vegetation survey

Plant species were investigated at two points (namely, Point A and Point B) in the waste dump in a coal mine in Indonesia in conjunction with literature research on plant species focusing on fast-growing characteristics and acid tolerance in order to understand the effects of acid soils on plant growth. Point A is about 400 m away from Point B in the same waste dump. This waste dump was constructed more than a year ago by piling up waste rocks, followed by revegetation which is mandatory for environmental conservation in mines. The revegetation in the research area had been conducted in the two stages (primary and secondary revegetation) in terms of growth rate of plants based on the experiences of the revegetation. Fast-growing trees were planted in the first stage of the revegetation for 3 years to improve soil conditions by increasing organic matter, followed by planting local plants in the second stage. In this mine, the death of plants has been reported along with the formation of acidic water at Point A as shown in **Figure 1**; on the other hand, it has not been reported at Point B.

In addition, three samples of *Eleusine indica (E. indica)* and *Melaleuca leucadendra (M. leucadendra) (Melaleuca cajuputi)* which were observed at the both points were taken, and they were separated into leaves, stems, and roots. The samples were washed with deionized water using a sonication (UT-106H, SHARP) at room temperature to remove soil particles. Finally, they were dried at 60°C for 72 hours and pulverized using mortar and pestle by each part of the plants. 0.25 g of each part of the samples were digested by 5 mL of a mixture of 61% nitric acid (HNO_3) and 35% hydrochloric acid (HCl) at a ratio of 3:1 at 110°C in a DigiPREP Jr. (SCP Science, Quebec, Canada) until they were completely digested with reference to the method by Quadir et al. [11]. In the case that they were not dissolved in the mixture, 1 mL of the mixture was added and the dissolution process was repeated. After the dissolution process, the volume of the solution was adjusted to 20 mL by adding deionized water. The solutions were subjected to Inductively Coupled Plasma-Atomic Emission Spectrometry (ICP-AES, VISTA-MPX ICP-OES [Seiko Inst., Japan]) after the filtration with 0.45 µm of membrane filter in order to quantify the content of Al, B, Fe, Mn, S, and Zn, which are thought to affect plant growth and be linked with the formation of acidic water, in each part of plant body. The results were calculated with mg per dry unit weight (mg/g) and compared with the waste water quality, which was recorded at Point A and Point B, so as to understand the linkage between the formation of acidic water and the accumulation of the elements in the plants.

2.2. Soil analysis

Soils within the waste dump were sampled until 100 cm depth with a 8-cm diameter hand auger (DIK-100A-55) at Point A and Point B in order to understand the current soil conditions. Soil pH was measured at each depth using Soil Acid meter (SK-910A-D, Sato, and DM-13, Takemura Electric Works Ltd.). The samples were separated by 20-30 cm and supplied to X-ray diffraction (XRD) and X-ray fluorescence (XRF) analysis, paste pH test, acid base accounting (ABA) test, and net acid generation (NAG) test so as to investigate the cause of acidic water in this area [12, 13] The XRD analysis was conducted using wide angle goniometer RINT 2100 XRD after the drying process at 50°C for 48 hours in a nitrogen atmosphere under the following conditions: radiation CuKα, operating voltage 40 kV, current 26 mA, divergence slid 1°, anti-scatter 1°, receiving slit 0.3 mm, step scanning 0.050°, scan speed 2.000°/min, and scan range 2.000–65.000°. In paste pH test, the change of pH was reported as paste pH after 12 hours of the dissolution process at the 1:2 of mixing ratio of sample and deionized water (pH$_{1:2w}$). The ABA and NAG test were performed to evaluate the acid producing potential of the soils. Net acid producing potential (NAPP) was calculated with acid potential (AP) and acid neutralizing capacity (ANC) of the samples as an acid producing potential in addition to NAG pH [12, 13]. Soils with NAG pH < 4.5 and NAPP > 0 were considered the source of acidic water, which can produce acids [14]. Acid producing potential generally rises with the increase of NAPP and the decrease of NAG pH.

2.3. Vegetation test

The vegetation test was conducted with a focus on the effect of acidic conditions on plant growth over time based on the results of the vegetation survey. Simulated acid soil was prepared by mixing sand, clay, and pyrite. Pyrite was mixed in the simulated soils, which were prepared using sand and clay in conformity to the soil texture and the physical properties of the topsoil in the mine site, aiming at setting the sulfur content as from 0 to 30% by weight on the basis of the result of XRF analysis. Each sample was labeled as S0.0, S0.5, S1.5, S3.0, S5.0, S7.5, S10.0, S15.0, and S20.0. The prepared acid soils were evenly mixed in each flower pot at a constant rate to uniform physical conditions in each pot, and used for the vegetation test. In this study, *A. mangium* which inhabits tropical forests in Southeast Asian countries, including Indonesia, and has acid tolerance was used: the seeds were obtained in Japan. *A. mangium* occurs naturally in the humid tropical lowland and has been successfully applied in reclamation in post-mine lands for bauxite, copper, coal, and iron in the world. It grows in compacted soils, dry area, on the slopes of hills, and humid area owing to high adaptability. *A. mangium* is designated an Al-excluder plant as well as *M. leucadendra* (*M. cajuputi*) [7].

In order to elucidate the effects of acidic conditions, including low pH, Al, B, Fe, Mn, S, and Zn on plant growth, *A. mangium* was planted on the prepared acid soils in the phytotron glass room G-9 in Biotron Application Center, Kyushu University under the following conditions: at 30°C room temperature and 70% relative humidity assuming the local climate in the mine site in Indonesia. The 9 flower pots with different content of pyrite were used for the vegetation test. In this test, five plants were planted in pots by the content of pyrite, and the height and the diameter were measured every week. About 500 mL of water was supplied to the pots every 3–4 days. The liquid fertilizer HYPONeX-R (N-P-K = 6-10-5) diluted to 1000 mg/L with water was, additionally, added to them once per week. The test was continued for 133 days until a clear distinction is observed.

To understand the effects of chemical and physical conditions of the prepared acid soils on plant growth, paste pH, NAG test, ABA test, Atterberg Limits test, and particle size distribution test were conducted according to the standard of AMIRA [13], ASTM D4318-05 [15], and ASTM D422-63 [16]. Moreover, the leachate from the bottom of the pots was taken every week to monitor the change of pH. At the end of the vegetation test, each part of the plants was digested by the acids in the same way as in Section 2.1 [11]. On the basis of the concentration of Al, B, Fe, Mn, S, and Zn in each part of the plants, the effect of the elements on plant growth over time was elucidated.

3. Results and discussion

3.1. Effects of acid soils on plant growth in mine site

Tables 1 and **2** summarize the main species of plants, which were planted in each stage of the revegetation in the mine and were characterized according to the literatures. The plants, which are acid tolerant, were planted in the first stage of the revegetation, and most of them were classified into fast-growing species. The plants generally planted for soil improvement, such as *Calliandra calothyrsus*, *Gliricidia sepium*, *Pterocarpus indicus*, and *Paraserianthes falcataria*, were planted in the first stage, aiming at improving soil conditions by increasing organic matter in the waste dump before planting local plants that are not acid tolerant in the second stage. Such fast-growing species shorten the time for revegetation and enable us to perform an earlier improvement of soil conditions in the waste dump. On the other hand, local plants which were planted in the second stage are utilized for industrial use as timber and ecosystem protection, such as *Intsia bijuga* and *Inocarpus fagifer*. The results indicated that plants were transplanted in the waste dump in the two stages for different purposes for successful revegetation in this mine.

The plants, which were observed in the waste dump at Point A and Point B, are summarized in **Table 3**. *Swietenia macrophylla*, *Mimosa pudica*, and cover crop (*Convolvulaceae*) plants, which are not acid tolerant, were not observed at Point A. Although *Intsia bijuga*, which is not acid tolerant, was observed at Point A, some of them withered. By contrast, *Paraserianthes falcataria*, and *M. leucadendra*, which are acid tolerant, were found at Point A and Point B, indicating that the plant growth in the waste dump may have been affected by acidic conditions. Additionally, the revegetation failed at Point A since *Swietenia macrophylla* which was planted in the second stage of the revegetation and the cover crop (*Convolvulaceae*) plants which are important for improving soil conditions withered.

Name of plants	Fast growing	Acid tolerance	References
Paraserianthes falcataria	+	+	Evans and Szott [17] Krisnawati et al. [18]
Pterocarpus indicus	−	+	Evans and Szott [17] Thomson [19]
Michelia champaca	−	+	Orwa et al. [20] Fern [21]
Gliricidia sepium	+	+	Orwa et al. [20] Evans and Szott [17]
Anthocephalus cadamba	+	+	Irawan and Purwanto [22] Krisnawati et al. [18]
Anthocephalus macrophyllus	+	+	Irawan and Purwanto [22] Krisnawati et al. [18]
Senna siamea	−	+	Orwa et al. [20]
Calliandra calothyrsus	+	+	Orwa et al. [20]
Melaleuca leucadendra	+	+	Masitah et al. [23] Turnbull [24] Nakabayashi et al. [25]
Nauclea orientalis	−	−	Orwa et al. [20]
Enterolobium cyclocarpum	+	+	National Research Council [26] Evans and Szott [17]

Table 1. Plant species in the first stage of the revegetation and the growth characteristics and the acid tolerance: + indicates that it has the characteristics; − indicates it does not.

Name of plants	Fast growing	Acid tolerance	References
Pterospermum javanicum	−	−	
Ficus benjamina	++	−	Fern [21]
Aquilaria spp.	−	−	Soehartono and Newton [27]
Inocarpus fagifer	+	+	Pauku [28]
Tectona grandis	++	−	Orwa et al. [20]
Hevea brasiliensis	−	−	Orwa et al. [20]
Pericopsis mooniana	−	−	Ishiguri et al. [29]
Swietenia macrophylla	−	−	Krisnawati et al. [18]
Shorea balangeran	−	−	
Intsia bijuga	−	−	Thaman et al. [30]
Schima wallichii	−	−	Orwa et al. [20]
Mimusops elengi	−	−	Kadam [31]
Fagraea fragrans	−	−	Steinmetz [32]
Samanea saman	++	−	National Research Council [26]

Table 2. Plant species in the second stage of the revegetation and the growth characteristics and the acid tolerance: ++ indicates that it has the characteristics; + indicates moderate; and − indicates that it does not.

Name of plants	Point A	Point B	Acid tolerance
Paraserianthes falcataria	O	O	+
Anthocephalus cadamba	O	O	+
Melaleuca leucadendra	O	O	+
Intsia bijuga	O	O	−
Eleusine indica	O	O	+
Swietenia macrophylla	×	O	−
Mimosa pudica	×	O	−
Cover crop (Convolvulaceae)	×	O	−

Table 3. Plant species, which were observed in the waste dump at Point A and Point B, and its acid tolerance: O indicates that it was observed; × indicates that it was not observed; + indicates acid tolerant; and − indicates that it is not acid tolerant.

In regard to waste water quality in the waste dump, electric conductivity (EC) and oxidation reduction potential (ORP) exhibited higher values at Point A (EC = 3.00 mS/cm, ORP = 588 mV) than at Point B (EC = 0.62 mS/cm, ORP = 568 mV). Moreover, a higher concentration of total Fe, SO_4^{2-}, and Al related to the formation of acidic water was recorded at Point A than at Point B, suggesting the formation of acidic water with the dissolution of sulfides at Point A based on the XRD results. This implies that acid soils can be formed with the intrusion of acidic water into the ground at Point A. While acid soils in Southeast Asian countries are often attributed to sulfuric sediments formed in mangrove, acidic water caused by the exposure of sulfides to oxygen and water with the excavation in mine site may have resulted in the formation of acid soils in this case [33]. The geochemical properties of the soils at Point A and Point B are summarized in **Table 4**. Paste pH showed similar values at each depth at the points. Furthermore, there was not a significant difference in soil pH showing 5.0–6.4 at each depth at the points. This was resulted from acid soils, which are common in the Southeast Asian countries. Meanwhile, there was a large difference in sulfur content, NAPP, and NAG pH between the points. Sulfur content showed larger values especially at 30–70 cm depth at Point A than at Point B. The results of XRD analysis suggested the presence of sulfides at the points, thus showing that the difference in the content of sulfides led to the difference in sulfur content at 30–70 cm depth between Point A and Point B. Since NAPP is calculated by subtracting ANC from AP, NAPP showed negative values at Point B where ANC showed positive values because of neutralization by carbonate and/or clay minerals. Besides, considering the similar values of paste pH and soil pH between Point A and Point B and complete dissolution of the samples with H_2O_2 in NAG test, the differences in NAG pH between the points were triggered by the abundance distribution of sulfides. The change of paste pH is not greatly affected by the dissolution of sulfides as soluble minerals mostly affect paste pH in a relative short time as well as soil pH. On the other hand, NAG pH significantly depends on the dissolution of sulfides owing to the complete dissolution of samples with H_2O_2 in NAG test. Therefore, NAG pH was lower at Point A than Point B since sulfides at Point A were sufficient to react with H_2O_2. With respect to the formation of acidic water, the presence of sulfides leads to the continuous formation of acidic water for a long time,

Point	Depth (cm)	Paste pH	S (%)	AP	ANC	NAPP (kg H$_2$SO$_4$/ton)	NAG pH
Point A	0–20	4.47	0.13	4.1	0.0	4.1	4.24
	30–50	4.72	0.75	23.1	0.0	23.1	3.76
	50–70	5.05	1.06	32.5	3.2	29.3	2.74
	70–100	4.41	0.52	16.0	0.0	16.0	3.22
Point B	0–20	4.74	0.09	2.9	43.9	−41.0	4.39
	30–50	4.76	0.08	2.5	43.6	−41.1	4.57
	50–70	4.57	0.11	3.4	43.4	−39.9	4.30
	70–100	4.63	0.11	3.3	43.4	−40.1	4.06

Table 4. Geochemical properties of the samples in the waste dump at Point A and Point B.

which is considered as a lag time due to the difference in the solubility of sulfur in various minerals, such as sulfates and sulfides [34]. The continuous dissolution of sulfides for a long time resulted in a high concentration of total Fe, Al, and SO$_4^{2-}$ in the waste water at Point A. Additionally, NAG pH < 4.5 and NAPP > 0 at Point A indicated the source of acidic water. This suggested that the evaluation of soil pH combined with NAPP and NAG pH, which are used to predict the formation of acidic water, enables us to understand the formation of acidic water and acid soils over time. Hence, in this case, the formation of acidic conditions in soils triggered by acidic water for a long time with the continuous dissolution of sulfides caused plant death at Point A. It is necessary to consider a lag time of the dissolution of sulfur in addition to soil pH for successful revegetation.

Figures 2 and **3** show the concentration of Al, As, B, Fe, Mn, S, and Zn in each part of *E. indica* and *M. leucadendra*, which were sampled in the waste dump at Point A and Point B. In addition, the standard deviation of the results was summarized by the species as shown in **Figure 4**. In particular, Fe and Al were accumulated in the roots of the plants. S was accumulated in the roots and leaves of the plants, and the concentrations of the elements were higher at Point A than that at Point B. This was attributed to the biological action to accumulate the excess of the harmful elements for plant growth on the roots. As the accumulation of Al in the roots causes the death of plants by preventing the absorption of nutrients from the roots, the high concentration of Al caused the inhibition of the growth of *Intsia bijuga*, *Swietenia macrophylla*, *Mimosa pudica*, and cover crop (*Convolvulaceae*) at Point A [35]. Moreover, a high concentration of Fe and S, which were derived from the dissolution of sulfides such as FeS$_2$, suggested that the dissolution of Al under acidic conditions was caused by the formation of acidic water with the dissolution of sulfides. The higher concentration of Fe, S, and Al was, additionally, obtained in *M. leucadendra*, which is acid tolerant, than that in *E. indica*, indicating that the accumulation capacity of the elements in the plant body depends on the species. Compared to the standard deviation of B, Mn, and Zn with that of Al, Fe, and S, the standard deviation of B, Mn, and Zn was near zero as shown in **Figure 4**, suggesting that B, Mn, and Zn were ubiquitous in the plants. B is an essential element for the maintenance of cell wall and carbohydrate metabolism [36], and the atomic number of B is similar with that of C, which is utilized for organic substances through the formation of carbon

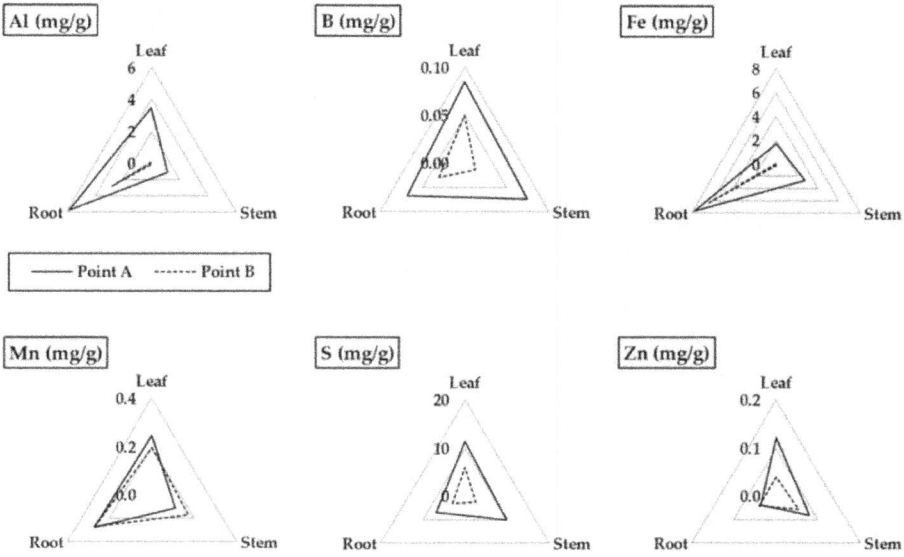

Figure 2. Concentration of Al, As, B, Fe, Mn, S, and Zn in each part of the plant body of *E. indica* at Point A and Point B.

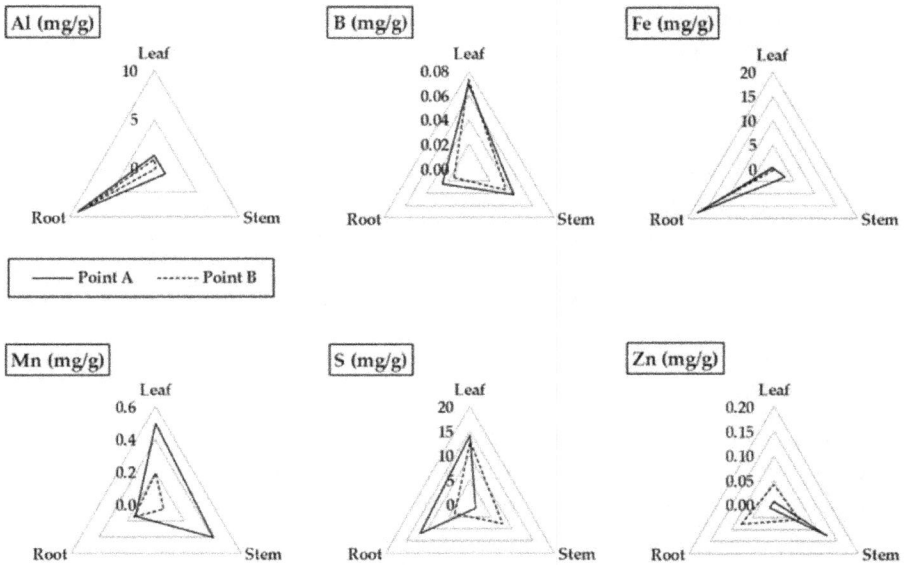

Figure 3. Concentration of Al, As, B, Fe, Mn, S, and Zn in each part of the plant body of *M. leucadendra* at Point A and Point B.

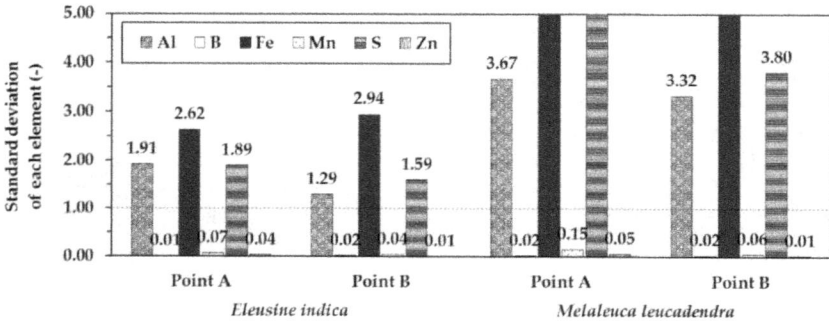

Figure 4. Standard deviation of the concentration of Al, As, B, Fe, Mn, S, and Zn in each part of the plant body: the distribution of the elements is homogeneous when standard deviations are nearly zero.

hydride in plant body. Thereby, B was widely distributed in the body of both plants in the similar way with C as a consequence of the transport mechanism of Mn to the leaves for photosynthesis [37, 38]. Zn is also one of the essential elements for plant growth as coenzyme to accelerate photosynthesis and DNA synthesis [39, 40]. Consequently, the dissolution of Al in acid soils triggered by the continuous formation of acidic water along with the dissolution of sulfides influenced the plant growth at Point A. For the presence of Al, the neutralization of soil conditions with limestones and/or chemicals is not always useful to improve soil conditions in acid soils because Al is released as $Al(OH)_4^-$ under alkali conditions. The evaluation of soil conditions before revegetation and/or farming is more important than the treatment of such acidic conditions. Likewise, even if the plants which are planted in the first stage of revegetation are acid tolerant, it is still necessary to select a proper plant for successful revegetation from the point of view of Al tolerance of plants and the dissolution of Al with the formation of acidic water over time in this case.

3.2. Tolerance characteristics of plants to acid soils

Figure 5(a) and **(b)** shows the relationship between plasticity index (I_p) and liquid limit (W_L), which shows soil conditions under different water content and particle size distribution of the prepared acid soils, respectively. Soil classification is closely associated with physical characteristics of soils such as particle size distribution, which affects plant growth [41]. All of the prepared acid soils were categorized as high W_L silt, and there were not significant differences in the particle size distribution among the samples as shown in **Figure 5**. This indicated that the growth of plants was not affected by the physical characteristics of the soil samples during the vegetation test.

The geochemical properties of the soil samples are summarized in **Table 5**, and **Figure 6** describes the content of Al, Fe, and S in the prepared acid soils. Besides, **Figure 7** shows the changes of the height of seedlings and the diameter of stem of *A. mangium* during the vegetation test. In **Table 5**, NAG pH dropped between S0.0 and S0.5, showing that NAG pH was significantly affected by the content of sulfides such as pyrite. The content of Fe and S rose with the increase in the mixing ratio of pyrite in **Figure 6**, whereas that of Al decreased, attributing to the decrease in the mixing ratio of simulated soils containing Al. In **Figure 7**,

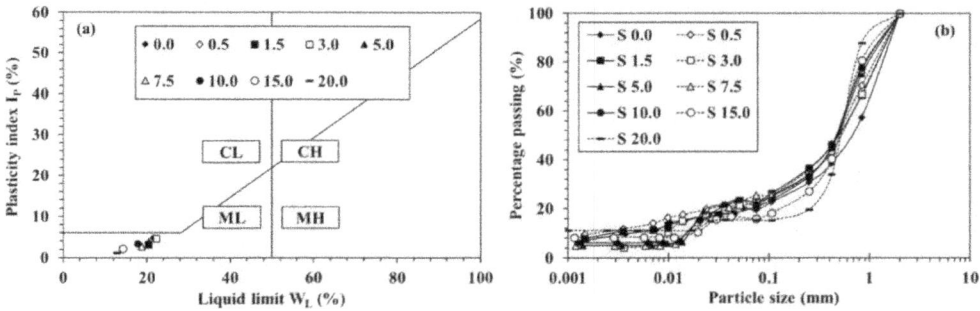

Figure 5. (a) Relationship between I_p and W_L of the prepared acid soils (C: clay; M: silt; H: high liquid limit; and L: low liquid limit). (b) Particle size distribution of the prepared acid soils.

the plants under the condition of S1.5–S20.0 withered after 56 days (8 weeks). By contrast, the plants under the condition of S0.0–S0.5 had grown after 56 days. For these results, the excess of sulfides in soils inhibited the growth of *A. mangium*, and the sulfur content for the limitation of the growth of *A. mangium* lied between 1.13 and 1.94% in this case. Furthermore, the plant growth declined with the increase in the mixing ratio of pyrite as indicated by the decrease of the height and the diameter as shown in **Figure 7**. The sulfur content at Point A showed 0.13–1.06%, especially 1.06% at 50–70 cm depth, as shown in **Table 4**, suggesting that the growth of plants which are subject to effects of acid soils compared to *A. mangium* can be inhibited under the soil conditions in the mine site.

In **Figure 8**, the changes of pH of the leachate from the bottom of the pots during the vegetation test are plotted. The pH of the leachate ranged from pH 2.0 to pH 4.0 in S1.5–S20.0 from the beginning of the experiment, resulting in the death of *A. mangium* contrary to the growth with constant pH 7.0 in S0.0. On the other hand, *A. mangium* in S0.5 had grown even if pH dropped

Sample	Sulfur content (%)	NAPP (kg H_2SO_4/ton)	NAG pH
S0.0	0.04	−8.9	5.67
S0.5	1.13	24.1	2.34
S1.5	1.94	48.8	2.24
S3.0	3.83	106.1	2.14
S5.0	6.86	198.2	2.05
S7.5	9.08	265.3	1.92
S10.0	11.60	340.1	1.98
S15.0	18.20	541.7	1.98
S20.0	28.20	846.0	1.91

Table 5. Geochemical properties of the prepared acid soils: the mixing ratio of pyrite in the prepared soils is labeled as sample names, e.g. the mixing ratio of pyrite is 0.5% in S0.5.

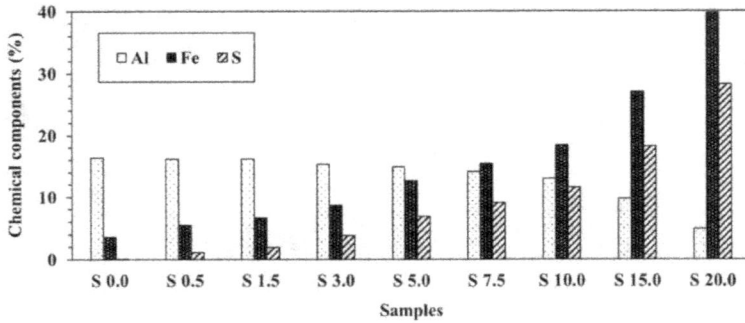

Figure 6. Content of Al, Fe, and S in the prepared acid soils with different contents of pyrite.

from pH 8.0 to ca. pH 2.0 after 70 days. It would appear that *A. mangium* can survive under acidic conditions by increasing the resistance against acidic conditions with the plant growth during 70 days [7]. The 8 cm height of seedlings were transplanted at the beginning of the experiment, and they died when they were exposed to pH 2.0 in S1.5–S20.0. However, *A. mangium* in S0.5 was exposed to pH 2.0 after 70 days when the height became 15 cm, leading to the existence of the seedlings due to the development of acid tolerance with the plant growth. Additionally, the sudden drop of the pH of the leachate may have resulted from the lag time of the dissolution of sulfides. Sulfides gradually dissolve in water over time, causing the formation of acidic water for a long time [34]. Thus, both formation of acidic water over time and the development of acid tolerance with the plant growth have to be taken into account for successful revegetation in the waste dump in mine site.

In **Figure 9**, the concentration of Al, As, B, Fe, Mn, S, and Zn in each part of *A. mangium* is summarized, and in **Figure 10**, the standard deviation of the results is described by the soil samples. Compared to the results in **Figures 2** and **3** and the concentration of the elements in *A. mangium* in **Figure 9**, Fe and Al were equally accumulated in the roots and S was accumulated in the roots

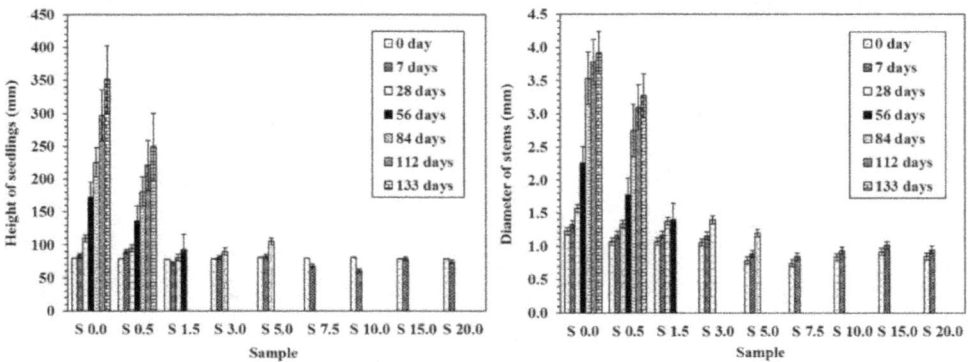

Figure 7. Change of the height and the stem diameter of *A. mangium* with different contents of pyrite in soils: the values were calculated based on the average of five samples by each content.

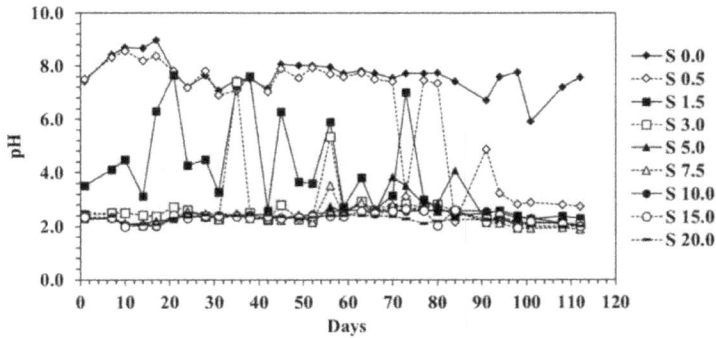

Figure 8. Change of pH of the leachate from the pots with different contents of pyrite in soils in the vegetation test.

and the leaves. Moreover, the accumulation of Al in the roots significantly rose with the increase in the mixing ratio of pyrite as shown in **Figure 9**. The high concentration of Al in the roots resulted in the death of *A. mangium* by preventing the absorption of nutrients from the roots [10]. In **Figure 10**, the standard deviation of Al, Fe, and S showed more than 0.5, whereas that of the other elements ranged from 0.01 to 0.1. The standard deviation of Al, Fe, and S, besides, rose with the increase in the mixing ratio of pyrite, revealing the accumulation of Al, Fe, and S in the plant body: the standard deviation of Al, Fe, and S was 2.98, 13.17, and 1.43 in S1.5, respectively. This was caused as a consequence of the biological action to accumulate the excess of the harmful elements for plant growth on the roots in the same case as in Section 3.1. The results in Section 3.1 also support that B, Mn, and Zn were distributed through the body of *A. mangium*. Furthermore, a larger amount of Fe, S, and Al was obtained in *M. leucadendra* compared to

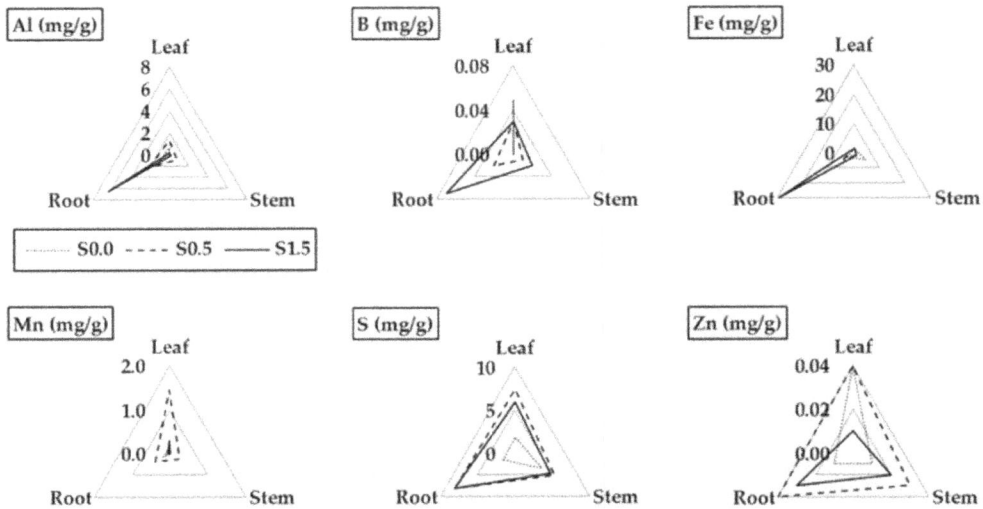

Figure 9. Concentration of Al, As, B, Fe, Mn, S, and Zn in each part of *A. mangium*.

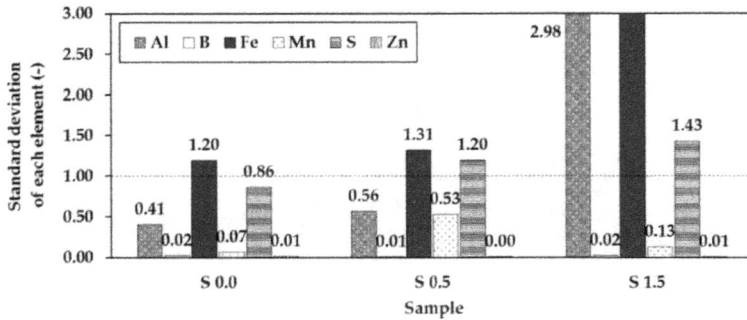

Figure 10. Standard deviation of the concentration of Al, As, B, Fe, Mn, S, and Zn in each part of *A. mangium*.

A. mangium although both *M. leucadendra* and *A. mangium* are acid tolerant. This indicated that the accumulation capacity of the elements in the plant body depends on the species.

Figure 11 demonstrates the growth of *A. mangium* in S0.0–S1.5, and **Figure 12** shows the change of height and length of the seedlings and roots of *A. mangium*. The number of the leaves and the height of *A. mangium* obviously decreased with the increase in the mixing ratio of pyrite as shown in **Figure 11**. The length of the roots also decreased with the increase of the content of pyrite as shown in **Figure 12**, attributing to the inhibition of the root elongation in response to Al-stress [10, 42]. This result and the increase of the content of Al in the roots of *A. mangium* with the increase of mixing ratio of pyrite revealed that the accumulation of Al in *A. mangium* resulted in the death in S1.5 regardless of the similar content of Al in S0.0–S1.5 as shown in **Figure 6**. In short, the immobilization of Al in soils without absorption in the plant body due to neutral pH led to the growth of *A. mangium* in S0.0, and *A. mangium* survived in S0.5 by increasing the resistance against acidic conditions with the growth and without absorption of Al in the plant body at around pH 7.0 during 70 days even if pH dropped in

Figure 11. Growth of *A. mangium* at different mixing ratios of pyrite in the acid soil.

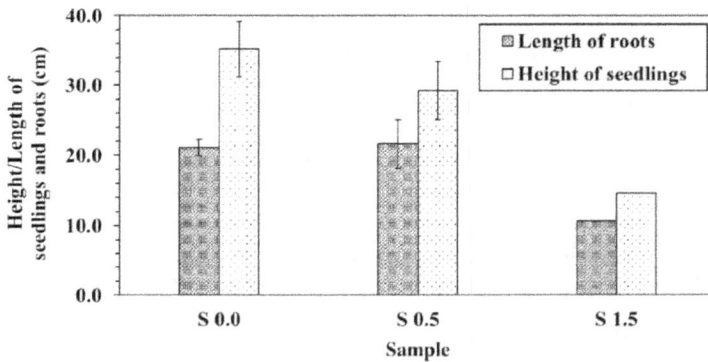

Figure 12. Change of the height of the seedlings and the length of the roots of *A. mangium* at the end of the vegetation test.

pH 2.0 after 70 days. In contrast, the accumulation of Al in the roots resulted in the inhibition of the root elongation and the death of *A. mangium* because of the large amount of dissolved Al and its accumulation in plants at pH 3.0 in S1.5. In S3.0–S20.0, *A. mangium* died resulting from the dissolution of Al and its accumulation in the roots with the continual production of H⁺ at pH 2.0 from the beginning of the vegetation test. Therefore, the timing of the transplant of plants and acidification of soils over time should be taken into account for the revegetation.

Acid-tolerant plants can survive in acid soils by setting up several tolerance mechanisms, such as the increase of soil pH around the root apices [5, 6]. However, the results in this study suggest that Al tolerance of plants has to be considered in addition to acid tolerance in the case that the accumulation of Al in plants inhibits plant growth. As shown in Ref. [7], there are differences in the Al-tolerant mechanism, such as Al-excluder and Al-accumulator. In regard to the effects of Al accumulation in plant body on plant growth, the plant classified as Al-excluder seems to be a suitable plant for the first stage of the revegetation in which acidic conditions are formed. Likewise, the content of Al was 6.45 mg/g in the roots of *A. mangium* in the vegetation test and it was 9.05 mg/g in *M. leucadendra,* which survived in acid soils in this study, although both plants are similarly categorized as Al-excluder in Ref. [7]. This indicates that *M. leucadendra* is more suitable for the first stage of the revegetation than *A. mangium* from the point of view of the effects of Al. Thereby, the plant for revegetation should be carefully selected from various perspectives, such as acid tolerance and Al tolerance of plants, and soil conditions. The evaluation in terms of not only soil conditions but also plant species has to be highlighted for successful revegetation.

4. Conclusions

In this study, vegetation survey and a vegetation test were conducted to investigate the current situation of acid soils and plant growth in mine site and to understand the effects of acid soils on plant growth over time. The results are summarized with the key to successful revegetation in terms of soil acidification and tolerance characteristics of plants as follows:

1. The dissolution of Al under acidic conditions in acid soils which were attributed to the formation of acidic water triggered by the dissolution of sulfides influenced plant growth in mine site. It is necessary to select a proper plant for successful revegetation from the point of view of Al tolerance and the dissolution of Al with the formation of acidic water over time.

2. Not only soil pH but also the assessment of the dissolution of sulfides over time is crucial for successful revegetation, suggesting that net acid producing potential (NAPP) and net acid generation (NAG) pH, which are used for evaluating the formation of acidic water, are useful to evaluate soil conditions for the revegetation in addition to soil pH.

3. The effects of acid soils on plant growth change according to plant species because Al-tolerant mechanism of plants depends on the species. Moreover, plants can survive under acidic conditions by increasing the resistance against acidic conditions with the plant growth. Therefore, the timing of the transplant of plants and acidification of soils over time should be taken into account for the revegetation.

Acknowledgements

The authors would like to express their appreciation to the mine for providing the samples. The experiments were conducted with the kind support of Mr. Shunta Ogata in Department of Earth Resources Engineering of Kyushu University.

Author details

Shinji Matsumoto[1]*, Hideki Shimada[2], Takashi Sasaoka[2], Ikuo Miyajima[3], Ginting J. Kusuma[4] and Rudy S. Gautama[4]

*Address all correspondence to: shin.matsumoto@aist.go.jp

1 Geological Survey of Japan, National Institute of Advanced Industrial Science and Technology (AIST), Ibaraki, Japan

2 Department of Earth Resources Engineering, Faculty of Engineering, Kyushu University, Fukuoka, Japan

3 Institute of Tropical Agriculture, Kyushu University, Fukuoka, Japan

4 Department of Mining Engineering, Institut Teknologi Bandung, Bandung, Indonesia

References

[1] Shishido M, Ito Y, Tamoto S. A vegetation method using native plants in the acid sulfate soil. Japan Society of Engineering Geology. 2013:123-124

[2] Ueno K. A mechanism of soil acidification in acid sulfate soils. Annual Report of Research Institute for Biological Function. 2004;**4**:25-33

[3] Kochian LV, Pineros MA, Hoekenga OA. The physiology, genetics and molecular biology of plant aluminum resistance and toxicity. Plant and Soil. 2005;**274**(1-2):175-195

[4] Matsumoto H. Plant responses to aluminum stress in acid soil molecular mechanism of aluminum injury and tolerance. Kagaku to Seibutsu. 2000;**38**(7):425-458

[5] Kochian LV, Hoekenga OA, Pineros MA. How do crop plants tolerate acid soils? Mechanism of aluminum tolerance and phosphorus efficiency. Annual Review of Plant Physiology and Plant Molecular Biology. 2004;**55**:459-493

[6] Vitorello VA, Capaldi FR, Stefanuto VA. Recent advances in aluminium toxicity and resistance in higher plants. Brazilian Journal of Plant Physiology. 2005;**17**:129-143

[7] Osaki M, Watanabe T, Tadano T. Beneficial effect of aluminum on growth of plants adapted to low pH soils. Soil Science & Plant Nutrition. 1997;**43**(3):551-563

[8] Tanikawa N, Inoue H, Nakayama M. Aluminum ions are involved in purple flower coloration in *Camellia japonica*. 'Sennen-fujimurasaki'. The Horticulture Journal. 2016;**85**(4):331-339

[9] Osaki M, Sittibush C, Nuyim T. Nutritional characteristics of wild plants grown in peat and acid sulfate soils distributed in Thailand and Malaysia. In: Vijarsorn P, Suzuki K, Kyuma K, Wada E, Nagano T, Takai Y, editors. A Tropical Swamp Forest Ecosystem and Its Greenhouse Gas Emission. Tokyo: Nodai Research Institute Tokyo University of Agriculture; 1995. p. 63-76

[10] Saifuddin M, Osman N, Idris RM, Halim A. The effects of pre-aluminum treatment on morphology and physiology of potential acidic slope plants. Kuwait Journal of Science and Engineering. 2016;**43**(2):199-220

[11] Quadir QF, Watanabe T, Chen Z, Osaki M, Shinano T. Ionomic response of *Lotus japonicus* to different root-zone temperatures. Soil Science and Plant Nutrition. 2011;**57**(2):221-232

[12] Sobek AA, Schuller WA, Freeman JR, Smith RM. Field and laboratory methods applicable to overburdens and minesoils. Report EPA-600/2-78-054. U.S. National Technical Information Service Report PB-280. 1978; 1-204. Available from: http://www.osmre.gov/resources/library/ghm/FieldLab.pdf [Accessed: Aug 31, 2017]

[13] AMIRA International. ARD Test Handbook: Prediction & Kinetic Control of Acid Mine Drainage, AMIRA P387A. Reported by Ian Wark Research Institute and Environmental Geochemistry International Ltd. Melbourne, Australia: AMIRA International; 2002. Available from: http://www.amira.com.au/documents/downloads/P387AProtocol Booklet.pdf [Accessed: Aug 31, 2017]

[14] Miller S, Robertson A, Donahue T. Advances in acid drainage prediction using the net acid generation (NAG) test. In: Proceedings of the 4th International Conference on Acid Rock Drainage; May 31, 1997–June 6, 1997; Vancouver, B.C., Canada: Natural Resources Canada; 1997. p. 533-549

[15]   ASTM. Standard Test Methods for Liquid Limit, Plastic Limit, and Plasticity Index of Soils ASTMD4318-05. Pennsylvania: ASTM International; 2005

[16]   ASTM. Standard Test Method for Particle-size Analysis of Soils (withdrawn 2016). ASTM D422-63(2007)e2. Pennsylvania: ASTM International; 2007

[17]   Evans DO, Szott LT, editors. Nitrogen fixing trees for acid soils. In: Proceedings of Nitrogen Fixing Tree Research Reports (EUA). Morrilton: Nitrogen Fixing Tree Association; Turrialba (Costa Rica): CATIE; 1994 328 p

[18]   Krisnawati H, Varis E, Kallio MH, Kanninen M. *Paraserianthes falcataria* (L.) Nielsen: Ecology, Silviculture and Productivity. Bogor: Center for International Forestry Research (CIFOR); 2011. 13 p. DOI: 10.17528/cifor/003394

[19]   Thomson LAJ. *Pterocarpus indicus* (narra), Ver. 2.1. In: Elevitch CR, editor. Species Profiles for Pacific Island Agroforestry. Holualoa: Permanent Agriculture Resources (PAR); 2006

[20]   Orwa C, Mutua A, Kindt R, Jamnadass R, Simons A. Agroforestree Database: A Tree Reference and Selection Guide, Ver. 4.0; 2009. Available from: http://www.worldagrofor-estry.org/treedb/AFTPDFS/Michelia_champaca.PDF [Accessed: Aug 31, 2017]

[21]   Fern K. *Michelia champaca*, Useful Tropical Plants Database; 2014. Available from: http://tropical.theferns.info/viewtropical.php?id=Magnolia+champaca [Accessed: Aug 31, 2017]

[22]   Irawan US, Purwanto E. White jabon (*Anthocephalus cadamba*) and red jabon (*Anthocephalus macrophyllus*) for community land rehabilitation: improving local propagation efforts. Agricultural Science. 2014;**2**(3):36-45

[23]   Masitah M, Shamsul Bahri AR, Jamilah MS, Ismail S. Histological observation of gelam (*Melaleuca cajuputi* Powell) in different ecosystems of terengganu. Journal of Biology Agriculture and Healthcare. 2014;**4**(26):1-7

[24]   Turnbull JW, editor. Multipurpose Australian Trees and Shrubs. Canberra: Australian Center for International Agricultural Research; 1986 316 p

[25]   Nakabayashi K, Nguyen NT, Thompson J, Fujita K. Effect of embankment on growth and mineral uptake of *Melaleuca cajuputi* Powell under acid sulphate soil conditions. Soil Science & Plant Nutrition. 2001;**47**(4):711-725

[26]   National Research Council (U.S.). Advisory Committee on Technology Innovation. Tropical Legumes: Resources for the Future: Report of an Ad Hoc Panel of the Advisory Committee on Technology Innovation, Board on Science and Technology for International Development, Commission on International Relat / National Research Council (U.S.). Advisory Committee on Technology Innovation. Washington: National Academy of Sciences; 1979 331 p

[27]   Soehartono T, Newton AC. Reproductive ecology of *Aquilaria spp.* in Indonesia. Forest Ecology and Management. 2001;**152**:59-71

[28]   Pauku RL. *Inocarpus fagifer* (Tahitian chestnut), Ver. 2.1. In: Elevitch CR, editor. Species Profiles for Pacific Island Agroforestry. Holualoa: Permanent Agriculture Resources (PAR); 2006

[29] Ishiguri F, Wahyudi I, Takeuchi M, Takashima Y, Iizuka K, Yokota S, Yoshizawa N. Wood properties of *Pericopsis mooniana* grown in a plantation in Indonesia. Journal of Wood Science. 2011;**57**(3):241-246

[30] Thaman RR, Thomson LAJ, DeMeo R, Areki R, Elevitch CR. *Instia bijuga* (vesi), Ver. 3.1. In: Elevitch CR, editor. Species Profiles for Pacific Island Agroforestry. Holualao: Permanent Agriculture Resources (PAR); 2006

[31] Kadam PV, Yadav KN, Deoda RS, Shivatare RS, Patil MJ. *Mimusops elengi*: A review on ethnobotany. Phytochemical and pharmacological profile. Journal of Pharmacognosy and Phytochemistry. 2012;**1**(3):64-74

[32] Steinmetz EF. *Fagraea fragrans*. Quarterly Journal of Crude Drug Research. 1961;**1**(2):66-71

[33] Dent DL, Pons LJ. A world perspective on acid sulphate soils. Geoderma. 1995;**67**(3-4): 263-276

[34] Matsumoto S, Shimada H, Sasaoka T, Matsui K, Kusuma GJ. Prevention of acid mine drainage (AMD) by using sulfur-bearing rocks for a cover layer in a dry cover system in view of the form of sulfur. Journal of the Polish Mineral Engineering Society. 2015;**2**(36):29-35

[35] Nguyen NT, Nakabayashi K, Thompson J, Fujita K. Role of exudation of organic acids and phosphate in aluminium tolerance of four tropical woody species. Tree Physiology. 2003;**23**(15):1041-1050

[36] Tanaka M, Miwa K, Fujiwara T. Molecular mechanism and regulation of boric acid transport in plants. The Journal of Japanese Biochemical Society. 2010;**82**(5):367-377

[37] Goulet RR, Pick FR. Changes in dissolved and total Fe and Mn in a young constructed wetland: Implications for retention performance. Ecological Engineering. 2001;**17**(4):373-384

[38] Sasaki K, Ogino T, Hori O, Takano K, Endo Y, Sakurai Y, Irie K. Treatment of heavy metals in a constructed wetland, Kaminokuni, Hokkaido: Accumulation of heavy metals in emergent vegetations. Journal of MMIJ. 2009;**125**(8):453-460

[39] Maeshima M. Zinc homeostasis and zinc signaling thoughts from plant zinc transporters. Seikagaku. 2014;**86**(3):407-410

[40] Kadlec RH, Knight RL. Treatment Wetlands. Boca Raton: CRC Press/Lewis Publishers; 1996

[41] Huang L, Dong BC, Xue W, Peng YK, Zhang MX, Yu FH. Soil particle heterogeneity affects the growth of a rhizomatous wetland plant. PLoS One. 2013;**8**(7):e69836

[42] Kikui S, Sasaki T, Maekawa M, Miyao A, Hirochika H, Matsumoto H, Yamamoto Y. Physiological and genetic analyses of aluminium tolerance in rice, focusing on root growth during germination. Journal of Inorganic Biochemistry. 2005;**99**(9):1837-1844

Soil Chemistry and Mineralogy

Fluoride Adsorption onto Soil Adsorbents: The Role of pH and Other Solution Parameters

Enos Wamalwa Wambu and Audre Jerop Kurui

Additional information is available at the end of the chapter

http://dx.doi.org/10.5772/intechopen.74652

Abstract

Soil adsorbents continue to attract increasingly high numbers of researchers in water defluoridation studies. An aspect of solution parameters, that is the aqueous adsorption of fluoride onto soil adsorbents in defluoridation studies, has been reviewed and reported. The pH was found to be the main factor controlling fluoride adsorption on the popular soil adsorbents including: aluminosilicates, iron (hydr)oxides, aluminum (hydr) oxides, apatites, carbonaceous minerals, calcareous soils and zeolites and the other key parameters being temperature, time of contact, and co-existent ions. Fluoride adsorption onto metal-exchanged zeolites and hydroxyapatites (optimum pH = 4–10), iron (hydro) oxide minerals (pH = 2–7), and carbonaceous minerals (pH = 4–12) is relatively pH-independent, and high amounts of fluoride are able to sorb upon the surfaces of these minerals in a wide range of pH values. However, montmorillonites (optimum pH = 5–6), aluminum (hydro)oxide minerals (pH = 5–7), and calcareous minerals (pH = 5–6) only sorb significant amount of fluoride in a narrow range of pH values. The fluoride adsorption onto the latter class of minerals, also generally occurring at slightly above room temperatures, appears to be highly specific and not strongly affected by the presence of coexistent anions including: PO_4^{3-}, SO_4^{2-}, Cl^-, and NO_3^-.

Keywords: adsorption, defluoridation, drinking water, fluoride, minerals, pH, soil

1. Introduction

Adequate dietary levels of fluoride are desired for good oral health and for the proper development of skeletal tissues [1]. Nonetheless, the excessive levels of fluoride in the environment pose major public health challenges in many regions of the world [2]. Dietary fluoride overexposure has been linked to a series of detrimental physiological effects [3] and it is known to lead to serious mottling of teeth enamel and gross skeletal malformations [4]. Continued dependence on fluoride-enriched water by communities in high-fluoride areas is the principal conduit by which people get exposed to undue levels of fluoride from the environment. Problems linked to prolonged consumption of excessive fluoride through water and food are, for that reason, normally correlated to areas of high-fluoride-bearing rocks and fluoride-enriched soil minerals. Even so, the hydro-geological release of soil mineral fluoride and its bioavailability through food chains is dependent on the hydrogeochemical characteristics of the environment.

Because of its detrimental public health effects when consumed in excessive amounts, the World Health Organization (WHO) has set recommended levels of fluoride for drinking water at 0.7 ppm [5]. However, the set maximum permissible levels of 1.5 ppm are the most widely used fluoride standards of drinking water to guard against dental caries and ensure healthy development of teeth and bones [6]. The point-of-use treatment of contaminated drinking water, to remove excessive fluoride while allowing sufficient levels for good oral and skeletal health, is now an indispensable component in many domestic water treatment protocols around the world [7]. Because of the high costs involved, many studies have recently been devoted to investigating the capacity of different materials for fluoride removal from water with a view to device more affordable approaches to water defluoridation [8–10].

Soil adsorbents, in particular, have been among natural media that have been extensively explored by researchers as alternate affordable media in water defluoridation [11–14] as they are normally more readily available and, by and large, possess significant fluoride adsorption capabilities. Furthermore, they are relatively stable and usable in a wider range of water conditions than most other natural media. The soil adsorbents that have attracted highest attention of scientists for water defluoridation include montmorillonites, aluminosilicates, iron and aluminum (hydr)oxides, hydroxyapatites, carbonaceous minerals, calcareous soils, and zeolites [15]. The solution pH, fluoride concentration, temperature, and co-existent ions play a major role in controlling fluoride adsorption onto soil adsorbents. Understanding the influence of these parameters in fluoride removal from water by adsorption using soil adsorbents could present additional insight into the scope of applicability of the geomaterials in water defluoridation.

The present work was initiated to interrogate available literature on water defluoridation by adsorption using soil adsorbents with a view to divulge information that could inform subsequent strategies in water defluoridation-based soil mineral adsorbents.

2. Adsorption surface enhancement for soil adsorbents

The potential soil adsorbents for fluoride sequestration from water are as diverse as the natural soil systems on earth. However, a glimpse through recent literature reveals that only few minerals have been repeatedly been studied for their potential to sorb fluoride from water over the last few decades. The selection of a soil adsorbent for water defluoridation studies is usually informed by, among other factors, the already known sorption capacities of the mineral for fluoride or for related adsorbates; the ease of availability of the mineral, its procurement, preparation, and applicability under given conditions; as well as by its user and environmental safety considerations. Based on approximate fluoride adsorption capacities of the minerals frequently revealed in the literature, the minerals that have exhibited the most promising potential for water defluoridation in the most recent studies include palygorskite (with a mean fluoride adsorption capacity of 57.97 mg/g), pumice (18.27 mg/g), zeolites (15.65 mg/g), hydroxyapatite (13.27 mg/g), iron-enriched laterites (9.39 mg/g), bauxite (7.53 mg/g), and montmorillonites (4.82 mg/g). The other minerals including kaolinites, ceramics, and quartz normally have mean fluoride adsorption capacities of less than 3.0 mg/g and do not constitute prospective robust fluoride adsorbents [15].

The capacity of soil media to sorb large amounts of fluoride is controlled by the predominant surface chemistry of the soil systems. The primary fluoride sorptive sites of clay colloids in the soil minerals comprise mainly the protonated or non-protonated silanol groups and the cationic positive centers provided by prevalent soil cations such as Fe^{3+}, Al^{3+}, and Si^{4+}. Natural soil systems are, however, generally associated with low ion-exchange capacities because the soil surfaces are normally saturated with replaceable counter groups, which mask and neutralize intrinsic surface charge so as to maintain mineral surface stability. The ion-exchange properties of the soil minerals can, however, be enhanced by pre-treatments that are aimed at dislodging the masking ions from the soil surfaces so as to increase the reactivity of the soil surfaces toward the target adsorbate ion and unblock the pores into the crystal lattice structure of the soil systems [16]. This is more so for soil surfaces that possess net charges that repel the adsorbate ions as is usually the case of fluoride adsorption onto clay systems, which are normally characterized by high density of electronegative oxygen groups in their structures that induce a net negative charge in the adsorbent soil surfaces. These surface charges make such soil to naturally repel and keep fluoride in the solution. This necessitates pretreatment to produce soil surface charge reversal in order to enhance their fluoride adsorption affinities and capacities.

The surface charge reversal for negatively charged soil adsorbents, which is aimed at enhancing electro-activity of their colloid surfaces towards aqueous fluoride particles, may be achieved by impregnating the adsorbent soil structure with multivalent metal ions or by grafting and intercalating the soil adsorbents with charged reactive groups. Hydrothermal activation of soil adsorbents in dilute acids is also a common practice that is not only simpler to apply but also more cost-effective [17, 18]. The latter procedure results in partial de-alumination of the clay structure, which increases the proportion of silica and the density of acid silanol groups on soil adsorbent surface leading to increased overall positive charge of the clay surfaces [19, 20].

3. Effect of selected solution parameters

The effects of adsorption solution parameters on the adsorption process spring from their influence on the adsorbents soil surface chemistry and on the flux transport of adsorbate solutes from the bulk solution through the aqueous matrix to the adsorbent surface. The principal solution parameters that control fluoride sequestration onto soil surfaces include the pH, temperature, contact time, fluoride concentration, and co-existing ions. Other contributing factors comprise: adsorbent dosage, adsorbent particle size, and the rate of agitation. The effect of adsorbent dosage and particle size and those of the adsorbate concentration mirrors each other. This is because both adsorbent dosage and particle size and those of the adsorbate concentration control the availability of reacting "particles" that drive the thermodynamic adsorption equilibrium on either side of the adsorption interface. Increase in the adsorbent dosage and in the adsorbate concentration results in high rates of adsorption as a result of more intensified solute fluxes through aqueous media to the soil surfaces. This influence is, however, extensively discussed elsewhere in the literature [15].

3.1. Effects of pH

Speciation and aqueous availability of fluoride in water is the function of pH, concentration, and the presence of cations such as: Al^{3+}, Fe^{3+}, Mn^{2+}, Ca^{2+}, and Mg^{2+} [21]. At low pH values of 4 and less, for example, the molecular HF fluoride species predominates aqueous fluoride speciation in solution. The formation of HF, which favors solubility and aqueous availability of fluoride, increases with decreasing pH of the media [22]. The fluoro-aluminum complexes that include AlF^{2+}, AlF_2^+, and AlF_3^0 and other metallo-fluoro complex species involving other multivalent cations such as Fe^{3+}, emerge in the pH range of 4–6 and the concentration of free fluoride ions in this pH range is only 21.35% [23]. At higher pH values, the stability of metallo-fluoro complexes decreases and the free fluoride anions, F^-, predominate. All fluorides exist as free anions, F^-, at pH values of 8–9, where all forms of aluminum species form the aluminate, $[Al(OH)_4]^-$, complexes in the presence of excess OH^- species [24].

In the same way, the solution pH controls the ionization of reactive surface groups in the colloid soil surfaces and determines the nature and the intensity of the soil surface charge and the adsorption potential at the soil surfaces [25]. Calcareous minerals, for instance, facilitate pH-dependent fluoride solubility according to the mass balance Eq. (1) as follows [26]:

$$CaCO_3(s) + H^+(aq) + 2F^-(aq) \rightleftharpoons CaF_2(s) + HCO_3^-(aq) \tag{1}$$

This equation relates calcite and fluorite in the natural soil environments when both salts are in contact with the water. Accordingly, the increase in pH and in the concentrations of HCO_3^- increases water fluoride concentrations and vice versa.

In addition, anionic adsorption onto soil adsorbents can proceed through specific or nonspecific adsorption. The former is based on ligand-exchange reactions where the anions displace OH^- and H_2O from the soil surfaces, whereas the latter involves electrostatic coulombic forces and mainly depends on the pH of zero net charge (pHpzc) of the adsorbent soil surface. Above pHzpc, the soil surface assumes positive charge, whereas below net positive surface charge persists [27]. The specific adsorption of fluoride by metal oxyhydroxide surface sites, for example, occurs by ligand exchange according to Eqs. (2) and (3) for protonated and nonprotonated sites, respectively, as follows:

$$SOH_2^+ \quad + \quad F^- \; \rightleftharpoons \; SF \quad + \quad H_2O \tag{2}$$

$$SOH \quad + \quad F^- \; \rightleftharpoons \; SF^- \quad + \quad H_2O \tag{3}$$

The pH of the aqueous media is, therefore, the prime factor that controls fluoride uptake by soil surfaces.

However, the solution pH of maximum fluoride adsorption varies from one type of soil adsorbent to the other. For iron-enriched laterites [27–29], kaolinites [22, 30–33] and, to a limited extent, for certain hydroxyapatites [34, 35], the maximum fluoride adsorption capacities occur in acidic media at pH values of 5 or less. Fluoride uptake in low pH (3–5) can be attributed to the formation of weak hydrofluoric acid [27]. It, therefore, shows that the adsorbent surfaces for these minerals have affinity for HF aqueous species.

The maximum fluoride adsorption capacities for montmorillonite clays [22, 36, 37], aluminum (hydrox)oxide minerals [38–44] and calcareous minerals [11, 12] are, however, restricted to pH values of 5–6. Montmorillonites, $Mx[(Mg, Al, Fe)_2(OH)_2(Si_4O_{10})].nH_2O$, are a group of expanding smectite clays comprising octahedral sheets of alumina sandwiched between two tetrahedral sheets of silica. The tripartite sheets are then loosely held together by weak oxygen-oxygen and oxygen-cation bonds [22]. The principal fluoride binding sides in montmorillonites are the cationic Fe^{3+}, Al^{3+} and Si^{4+} centers. At pH of 4 and less, the capacity of montmorillonites to sorb large amounts of fluoride is greatly compromised due to their disruptive dissolution of the mineral structure with release of Fe^{3+}, Al^{3+} and SiO_2. A major part of fluoride in a montmorillonite-water system exists in the form of aqueous iron and aluminum complexes, and only a small fraction is able to sorb onto the clay surface.

Conversely, certain soil sorbents, which include pumice [45, 46]; palygorskites [47]; and particular ferric oxide minerals such as hematite [48, 49] are able to sorb high amounts of fluoride over an entire range of pH values from 2 to about 8. Furthermore, fluoride adsorption onto natural and metal-exchanged zeolites [50] and onto a class of carbonaceous adsorbents including lignite [51, 52] and coal [52, 53] appear to be quite pH-independent and high amounts of fluoride adsorption based on this class of adsorbents occur over the wide range of pH values of 4–12.

In general, therefore, montmorillonites normally tend to solubilize in low pH media and get poisoned by excessive OH^- ions in alkaline media. For this reason, montmorillonites usually have narrow fluoride sorption edge within the neutral pH values. Like for montmorillonites, the usual pH for effective fluoride removal from water using metal-enhanced palygorskite is usually in the range of 2–8. Fluoride adsorption onto metal-exchanged zeolites and onto certain synthetic hydroxyapatites is, however, relatively independent of pH, and the adsorbents are able to take up high fluoride adsorption over a wide choice of pH values of 4–10. Aluminum oxide minerals usually have a narrower fluoride sorption edge in the pH range of 5.5–6.5 as is maximum fluoride adsorption onto Ca-based minerals, which occurs within the pH values of 5–6. On the other hand, high fluoride uptakes by hematite occur over a wider range of acidic pH values of 2–7. In the same way, optimum fluoride removal using carbonaceous adsorbents can be achieved at room temperature in the pH range of 4–12.

Differences in pH of maximum fluoride uptake for various soil systems arise principally from the differences in the surface chemistry of the mineral adsorbents, which control the affinity of soil surfaces towards different fluoride species in soil surfaces. It can be assumed that soils that have high fluoride adsorptions in strongly acidic media of pH 5 or less have higher affinity for molecular HF species, which are dominant in this range of medium pH. The HF particles adsorb by forming continuous hydrogen bonds with electronegative centers in the soil surfaces. Certain soils that preferentially sorb fluoride in the near-neutral acidic pH values of 5–6 have affinity for F^- species, and the mode of fluoride adsorption is mainly complexation with positive cationic centers in the soil colloid structure, which include Al^{3+}, Fe^{3+} and Si^{4+} among others. Soil adsorbent that sorb high fluoride levels over a wide range of pH values contains heterogeneous surfaces, which have attraction to several different fluoride species in solution.

3.2. Adsorption temperature

The effect of temperature on fluoride adsorption onto soil surfaces arise from its influence on the adsorption energy balance, on the kinetics of adsorbate particles, and on the chemical activation of reacting species. Higher temperatures enhance increased rates of adsorption by enhancement of faster solute transport from the bulk solution towards the adsorbent surfaces. Higher temperatures also raise the average energy of the particles allowing a higher number of particles to attain necessary activation energy to enable them to react. Very high temperatures may, however, counter the adsorption fluxes leading to reduced rates and magnitude of uptake of the adsorbate by the adsorbents.

As for the effects of solution pH, however, the effect of temperature on fluoride adsorption on popular soil adsorbents is varied. The peak fluoride adsorption by natural montmorillonites [22, 36, 37, 54], Fe(III)-modified montmorillonite [55], pumice [56] and lignite [51, 52] occur within a range of temperatures close to room temperature (298 K). Nevertheless, the highest fluoride uptake by both aniline-modified montmorillonites and pyrole-modified montmorillonites [57] as well as by coal [52, 53] is favored by above room temperatures close to 303 K. It has been found that fluoride-exchange reactions for hydroxyapatites [58–60] and for certain ferric oxide minerals such as hematite [48, 49] can occur over a wide range of temperatures of 298–323 K. Fluoride adsorption onto Mg^{2+} and Al^{3+} [47], Fe^{3+} [61] and ZrO^{2+} [62] loaded palygorskite minerals; synthetic hydroxyapatites [63, 64]; calcareous minerals [65] and onto magnesia-loaded fly ash cenospheres (MLC) is favored by higher temperatures in the range of 303–323 K. This indicates the existence of endothermic chemical surface reactions. The efficacies of bauxite to sorb fluoride has, however, been found to decrease with increasing temperature indicating the existence of exothermic fluoride immobilization in bauxite surfaces [39, 44].

3.3. Contact time

The resident time required for equilibration in an adsorption process depends mainly on the adsorbent structure and on the nature of reactions that occur between the adsorbate particles and reactive sites at the adsorbent surface. Adsorbents with compact crystalline structures and characteristically surface exposed reactive sites tend to have rapid rates of adsorption than porous media with intraparticle sorptive sites. This is because in the latter case, the adsorbate particles have to be transported by diffusion into the inner adsorbent structures in order to access the reactive adsorbent sites. Fluoride adsorption onto pulverized crystalline calcareous minerals tends to occur rapidly by surface precipitation of fluorite, CaF_2, over the entire mineral surface [66, 67]. Water defluoridation using calcareous materials is, as a result, normally characterized by fast adsorption rates and the adsorption equilibrium lies within the range of 30–60 min [11, 68].

In less crystalline adsorbents such as lignite, more than 90% fluoride adsorption occurs within the initial 10 min. It, however, takes up to 150 min to saturate the less exposed sites inside the adsorbent structure with the latter 10% of the process [51, 52]. Such trends are also observed in the case of fluoride adsorption onto coal with shorter equilibration periods of 60–90 min for the latter phase of adsorption [52, 53], which shows that coal is more crystalline and less porous than lignite.

Equilibration periods required for fluoride adsorption onto pumice have been reported to lie within the range of 20–30 min but pumice adsorbents have not generally been associated with the two-phase adsorption phenomenon. This indicates the presence of limited porosity in the mineral structure of these adsorbents [46, 56]. Although some authors have linked fluoride adsorption onto natural montmorillonites to rapid sorption rates associated with the short adsorption equilibrium periods of just 20–30 min [22, 37], several natural montmorillonites [36, 54] and Fe(III)-modified montmorillonites [55] appear to have consistent fluoride adsorption equilibrium time intervals in the range of 110–180 min. In the same way, a number of mineral adsorbents including metal-intercalated palygorskites [47] and certain aluminum oxide minerals [39, 44] appear to have equilibrium intervals within the same range of periods. This signifies that these minerals possess structural porosities that are comparable.

As in the case of fluoride adsorption onto calcareous and carbonaceous soil adsorbents, the immobilization of fluoride into adsorbent zeolites [50, 69, 70], hydroxyapatites [60, 63, 64], iron oxide minerals [49] as well as into certain classes of aluminum oxide minerals [41] is characterized by initial rapid phases of adsorption characterized typically by short equilibration intervals of just 10–30 min, which are then followed by prolonged equilibration that could extend to 10–48 h. The final slow phase of equilibration can be ascribed to high structural porosity as in the case of zeolites or to slow valence exchange reaction mechanisms characteristic of fluoride immobilization upon hydroxyapatite, hydr(oxide) aluminum and iron minerals.

3.4. Co-existent ions

Natural water systems contain dissolved species across the organic-inorganic chemical continuum. Co-existent ions in water control the adsorption of fluoride by their competitive effect for the sorptive space on the adsorbent soil surfaces and by their influence on the adsorbate flux from the bulk solution to the sorbent surface. Co-ions tend to lower the rates and magnitude of adsorption, but the extent of these influence largely depend on the chemical and geometric dimensions of the ions, relative concentrations and affinities of the individual ions for the adsorbent surface. The influence of interfering ion, however, varies from one soil adsorbent to the other.

The soil adsorbents whose fluoride uptake is most affected by co-existent anions include iron oxide minerals [27–29] and certain carbonaceous mineral adsorbents. The suppression of fluoride immobilization upon ferric oxide minerals in the presence of common anions follows the order: $PO_4^{3-} > SO_4^{2-} > Cl^- > NO_3^-$ [29]. Fluoride adsorption onto zeolites [50], HAps [59, 63, 64], bauxite [39, 40] and calcareous mineral adsorbents [11, 12, 58, 66, 67, 71, 72] is, however, site specific, and it is not normally affected by competing anions in solution. For that reason, the adsorbents are able to sorb relatively high amounts of fluoride independent of co-existing anions such as Cl^- NO_3^-, SO_4^{2-}, CH_3COO^- and PO_4^{3-} ions.

4. Conclusions

Soil adsorbents that have attracted the highest interest as possible adsorbents for the removal of fluoride from water include: aluminosilicates, iron and aluminum (hydr)oxides, apatites, carbonaceous minerals, calcareous soils and zeolites. It is found that the pH is the main solution factor controlling fluoride adsorption onto soil surface. The other contributing parameters include temperature, time of contact and co-existent ions. The montmorillonite clays, generally, solubilize in low pH media and get poisoned by excess OH^- ions in alkaline media. They are generally characterized by small fluoride sorption edge within the neutral pH values. The usual pH for efficient fluoride removal from water using metal-enhanced palygorskite is in the range of 32–2. Fluoride adsorption onto metal-exchanged zeolites and onto synthetic HAps is, however, independent of pH, and high fluoride adsorption occurs in the pH range of 4–10. Aluminum oxide minerals, on the other hand, usually have a narrow sorption edge in the pH range of 5.5–6.5. In the same way, maximum fluoride adsorptions onto most of the calcareous minerals occur within the pH values of 5–6. High fluoride uptakes by hematite occur over a wide range of pH (2–7) but optimum fluoride removal using carbonaceous adsorbents can be achieved at room temperature in the pH range of 4–12.

Author details

Enos Wamalwa Wambu* and Audre Jerop Kurui

*Address all correspondence to: wambuenos@yahoo.com

Department of Chemistry and Biochemistry, University of Eldoret, Eldoret, Kenya

References

[1] World Health Organisation. Writing Oral Health Policy: A Manual for Oral Health Managers in the WHO African Region. Brazzavile: WHO Regional Office for Africa; 2005

[2]   Ayoob S, Gupta K. Fluoride in drinking water: A review on the status and stress effects. Critical Reviews in Environmental Science and Technology. 2006;**36**:433-487. DOI: 10.1080/10643380600678112

[3]   Kalisinska E, Natalia IB, Bird ÁBÁ. Fluoride concentrations in the pineal gland, brain and bone of goosander (Mergus merganser) and its prey in Odra River estuary in Poland. Environmental Geochemistry and Health. 2014;**36**:1063-1077. DOI: 10.1007/s10653-014-9615-6

[4]   Kebede A, Retta N, Abuye C, Whiting SJ, Kassaw M, Zeru T, et al. Dietary fluoride intake and associated skeletal and dental fluorosis in school age children in rural Ethiopian Rift Valley. International Journal of Environmental Research and Public Health. 2016;**13**:756-766. DOI: 10.3390/ijerph13080756

[5]   Centre for Disease Control and Prevention (CDC). Private well water and fluoride. Priv. Well Water Fluoride. 2005:7-9. http://www.cdc.gov/fluoridation/fact_sheets/wellwater.htm

[6]   Gordon B, Callan P, Vickers C. WHO guidelines for drinking-water quality. 2008. DOI: 10.1016/S1462-0758(00)00006-6

[7]   Wimalawansa SJ. Purification of contaminated water with reverse Osmosis: Effective solution of providing clean water for human needs in developing countries. International Journal of Emerging Technology and Advanced Engineering. 2013;**3**(12):75-89

[8]   Sivarajasekar N, Paramasivan T, Muthusaravanan S, Muthukumaran P, Sivamani S. Defluoridation of water using adsorbents—A concise review. Journal of Environment & Biotechnology Research. 2017;**6**:186-198

[9]   Waghmare SS, Arfin T. Defluoridation by adsorption with chitin-chitosan-alginate-polymers-cellulose-resins-algae and fungi—A review. International Research Journal of Engineering and Technology. 2015;**2**:1178-1197

[10] Loganathan P, Vigneswaran S, Kandasamy J, Naidu R. Defluoridation of drinking water using adsorption processes. Journal of Hazardous Materials. 2013;**248-249**:1-19. DOI: 10.1016/j.jhazmat.2012.12.043

[11] El-Said GF, Draz SEO. Physicochemical and geochemical characteristics of raw marine sediment used in fluoride removal. Journal of Environmental Science and Health. Part A, Toxic/Hazardous Substances & Environmental Engineering. 2010;**45**:1601-1615. DOI: 10.1080/10934529.2010.506117

[12] Nath SK, Dutta RK. Enhancement of limestone defluoridation of water by acetic and citric acids in fixed bed reactor. CLEAN—Soil, Air, Water. 2010;**38**:614-622. DOI: 10.1002/clen.200900209

[13] Goromo K, Zewgw F, Hundhammer B, Megersa N. Fluoride removal by adsorption on thermally treated lateritic soils. Bulletin of the Chemical Society of Ethiopia. 2012;**26**:361-372

[14] Hyun S, Kang DH, Kim J, Kim M, Kim D-Y. Adsorptive removal of aqueous fluoride by liner minerals from SPL-landfill leachate during the seepage process. Desalination. 2011;276:347-351. DOI: 10.1016/j.desal.2011.03.075

[15] Wambu EW, Ambusso W, Onindo CO, Muthakia GK, Wambu EW. Review of fluoride removal from water by adsorption using soil adsorbents—An evaluation of the status. Journal of Water Reuse and Desalination. 2016;6:1-29. DOI: 10.2166/wrd.2015.073

[16] Girgis BAY. Reuse of discarded deactivated bleaching earths in the bleaching of oils. Grasas Y Aceites. 2005;56:34-45

[17] Falaras P, Lezdou F, Seiragakis G, Petrakis D. Bleaching properties of alumina-pillared acid-activated Montemorillonite. Clays and Clay Minerals. 2000;48:549-556

[18] Wu Z, Li C, Sun X, Xu X, Dai B, Li J, et al. Characterization, acid activation and bleaching performance of Bentonite from Xinjiang. Chinese Journal of Chemical. Engineering. 2006;14:253-258

[19] Frini-Srasra N, Srasra E. Acid treatment of south Tunisian palygorskite: Removal of Cd(II) from aqueous and phosphoric acid solutions. Desalination. 2010;250:26-34. DOI: 10.1016/j.desal.2009.01.043

[20] Makhoukhi BB, Didi M, Villemin D, Azzouz A. Acid activation of Bentonite for use as a vegetable oil bleaching agent. Grasas Y Aceites. 2009;60:343-349. DOI: 10.3989/gya.108408

[21] Erdemoglu SB, Türkdemir H, Gücer S. Determination of total and fluoride bound aluminium in tea infusions by ion selective electrode and flame atomic absorption spectrometry. Analytical Letters. 2000;33:1513-1529. DOI: 10.1080/00032710008543140

[22] Agarwal M, Rai K, Srivastav R, Dass S. Fluoride speciation in aqueous suspensions of montmorillonite and kaolinite. Toxicological and Environmental Chemistry. 2002;82:11-21

[23] Jackson PJ, Harvey PW, Young WF. Chemistry and Bioavailability Aspects of Fluoride in Drinking Water. Marlow, Bucks: Medmenham; 2002

[24] Richards LA, Vuachère M, Schäfer AI. Impact of pH on the removal of fluoride, nitrate and boron by nanofiltration/reverse osmosis. Desalination. 2010;261:331-337. DOI: 10.1016/j.desal.2010.06.025

[25] Zhu M, Ding K, Jiang X, Wang H. Investigation on co-sorption and desorption of fluoride and phosphate in a red soil of China. Water, Air, and Soil Pollution. 2007;183:455-465. DOI: 10.1007/s11270-007-9394-0

[26] Jha SK, Singh RK, Damodaran T, Mishra VK, Sharma DK, Rai D. Fluoride in groundwater: Toxicological exposure and remedies. Journal of Toxicology and Environmental Health. Part B, Critical Reviews. 2013;16:52-66. DOI: 10.1080/10937404.2013.769420

[27] Sujana MG, Pradhan HK, Anand S. Studies on sorption of some geomaterials for fluoride removal from aqueous solutions. Journal of Hazardous Materials. 2009;161:120-125. DOI: 10.1016/j.jhazmat.2008.03.062

[28] Maiti A, Basu JK, De S. Chemical treated laterite as promising fluoride adsorbent for aqueous system and kinetic modeling. Desalination. 2011;**265**:28-36. DOI: 10.1016/j. desal.2010.07.026

[29] Huang Y, Shih Y-J, Chang C-C. Adsorption of fluoride by waste iron oxide: The effects of solution pH, major coexisting anions, and adsorbent calcination temperature. Journal of Hazardous Materials. 2011;**186**:1355-1359. DOI: 10.1016/j.jhazmat.2010.12.025

[30] Sugita H, Komai T, Okita S, Tokanaga S, Matsunaga I. Analysis on adsorption behavior of fluorine on clay minerals using Freundlich. Journal of the Mining and Materials Processing Institute of Japan. 2005;**121**:416-422

[31] Gogoi PK, Baruah R. Fluoride removal from water by adsorption on acid activated kaolinite clay. Indian Journal of Chemical Technology. 2008;**15**:500-503

[32] Sundaram SC, Viswanathan N, Meenakshi S. Fluoride sorption by nano-hydroxyapatite/chitin composite. Journal of Hazardous Materials. 2009;**172**:147-151. DOI: 10.1016/j. jhazmat.2009.06.152

[33] Wei S, Xiang W. Surface properties and adsorption characteristics for fluoride of kaolinite, ferrihydrite and kaolinite-ferrihydrite association. Journal of Food, Agriculture and Environment. 2012;**10**:923-929

[34] Gao S, Sun R, Wei Z, Zhao H, Li H, Hu F. Size-dependent defluoridation properties of synthetic hydroxyapatite. Journal of Fluorine Chemistry. 2009;**130**:550-556. DOI: 10.1016/ j.jfluchem.2009.03.007

[35] Mourabet M, El Rhilassi A, El Boujaady H, Bennani-Ziatni M, El Hamri R, Taitai A. Removal of fluoride from aqueous solution by adsorption on Apatitic tricalcium phosphate using Box–Behnken design and desirability function. Applied Surface Science. 2012;**258**:4402-4410. DOI: 10.1016/j.apsusc.2011.12.125

[36] Tor A. Removal of fluoride from an aqueous solution by using montmorillonite. Desalination. 2006;**201**:267-276. DOI: 10.1016/j.desal.2006.06.003

[37] Ramdani A, Taleb S, Benghalem A, Ghaffour N. Removal of excess fluoride ions from Saharan brackish water by adsorption on natural materials. Desalination. 2010;**250**: 408-413. DOI: 10.1016/j.desal.2009.09.066

[38] Farrah H, Slavek J, Pickering WF. Fluoride interactions with hydrous aluminum oxides and alumina. Australian Journal of Soil Research. 1987;**25**:55-69

[39] Sujana MG, Thakur RS, Rao SB. Removal of fluoride from aqueous solution by using alum sludge. Journal of Colloid and Interface Science. 1998;**206**:94-101. DOI: 10.1006/ jcis.1998.5611

[40] Das N, Pattanaik P, Das R. Defluoridation of drinking water using activated titanium rich bauxite. Journal of Colloid and Interface Science. 2005;**292**:1-10. DOI: 10.1016/j. jcis.2005.06.045

[41] Jiménez-Becerril J, Solache-Ríos M, García-Sosa I. Fluoride removal from aqueous solutions by Boehmite. Water, Air, & Soil Pollution. 2012;**223**:1073-1078. DOI: 10.1007/s11270-011-0925-3

[42] Lavecchia R, Medici F, Piga L, Rinaldi G. Fluoride removal from water by adsorption on a high alumina content bauxite. Chemical Engineering Transactions. 2012;**26**:225-230

[43] Sujana MG, Anand S. Iron and aluminium based mixed hydroxides: A novel sorbent for fluoride removal from aqueous solutions. Applied Surface Science. 2010;**256**:6956-6962. DOI: 10.1016/j.apsusc.2010.05.006

[44] Mohapatra D, Mishra D, Mishra SP, Chaudhury GR, Das RP. Use of oxide minerals to abate fluoride from water. Journal of Colloid and Interface Science. 2004;**275**:355-359. DOI: 10.1016/j.jcis.2004.02.051

[45] Malakootian M, Moosazadeh M, Yousefi N, Fatehizadeh A. Fluoride removal from aqueous solution by pumice: Case study on Kuhbonan water, African. Journal of Environmental Science and Technology. 2011;**5**:299-306

[46] Asgari G, Roshani B, Ghanizadeh G. The investigation of kinetic and isotherm of fluoride adsorption onto functionalize pumice stone. Journal of Hazardous Materials. 2012;**217-218**:123-132. DOI: 10.1016/j.jhazmat.2012.03.003

[47] Zhang J, Xie S, Ho Y. Removal of fluoride ions from aqueous solution using modified attapulgite as adsorbent. Journal of Hazardous Materials. 2009;**165**:218-222. DOI: 10.1016/j.jhazmat.2008.09.098

[48] Mohapatra M, Padhi T, Anand S, Mishra BK. CTAB mediated Mg-doped nano Fe_2O_3: Synthesis, characterization, and fluoride adsorption behavior. Desalination and Water Treatment. 2012;**50**:376-386. DOI: 10.1080/19443994.2012.720411

[49] Sequeira A, Solache-Ríos M, Balderas-Hernández P. Modification effects of hematite with aluminum hydroxide on the removal of fluoride ions from water. Water, Air, & Soil Pollution. 2011;**223**:319-327. DOI: 10.1007/s11270-011-0860-3

[50] Xu YH, Ohki A, Maeda S. Removal of arsenate , phosphate, and fluoride ions by aluminium-loaded shirasu-zeolite. Toxicological and Environmental Chemistry. 2000;**76**:111-124

[51] Pekař M. Fluoride anion binding by natural lignite (south Moravian deposit of Vienna Basin). Water, Air, and Soil Pollution. 2008;**197**:303-312. DOI: 10.1007/s11270-008-9812-y

[52] Sivasamy A, Singh KP, Mohan D, Maruthamuthu M. Studies on defluoridation of water by coal-based sorbents. Journal of Chemical Technology and Biotechnology. 2001;**76**:717-722. DOI: 10.1002/jctb.440

[53] Borah L, Dey NC. Removal of fluoride from low TDS water using low grade coal, Indian J. Chemical Technology. 2009;**16**:361-363

[54] Achour S, Youcef L. Defluoridation of the Algerian north Sahara waters by adsorption onto local bentonites. International Journal of Environmental Studies. 2009;**66**:151-165. DOI: 10.1080/00207230902859747

[55] Bia G, De Pauli CP, Borgnino L. The role of Fe(III) modified montmorillonite on fluoride mobility: Adsorption experiments and competition with phosphate. Journal of Environmental Management. 2012;**100**:1-9. DOI: 10.1016/j.jenvman.2012.01.019

[56] Mahvi AH, Heibati B, Mesdaghinia A, Yari AR. Fluoride adsorption by pumice from aqueous solutions. E-Journal of Chemistry. 2012;**9**:1843-1853. DOI: 10.1155/2012/581459

[57] Karthikeyan M, Kumar KKS, Elango KP. Studies on the defluoridation of water using conducting polymer/montmorillonite composites. Environmental Technology. 2012;**33**: 733-739

[58] Fan X, Parker DJ, Smith MD. Adsorption kinetics of fluoride on low cost materials. Water Research. 2003;**37**:4929-4937. DOI: 10.1016/j.watres.2003.08.014

[59] Mohapatra M, Anand S, Mishra BK, Giles DE, Singh P. Review of fluoride removal from drinking water. Journal of Environmental Management. 2009;**91**:67-77. DOI: 10.1016/j.jenvman.2009.08.015

[60] Nie Y, Hu C, Kong C. Enhanced fluoride adsorption using Al (III) modified calcium hydroxyapatite. Journal of Hazardous Materials. 2012;**233-234**:194-199. DOI: 10.1016/j.jhazmat.2012.07.020

[61] He ZL, Zhang GK, Xu W. Enhanced adsorption of fluoride from aqueous solution using an iron-modified Attapulgite adsorbent. Water Environment Research. 2013;**85**:167-174. DOI: 10.2175/106143012X13560205144218

[62] Zhang G, He Z, Xu W. A low-cost and high efficient zirconium-modified-Na-attapulgite adsorbent for fluoride removal from aqueous solutions. Chemical Engineering Journal. 2012;**183**:315-324

[63] Murutu CS, Onyango MS, Ochieng A, Otieno FO. Investigation on limestone-derived apatite as a potential low cost adsorbent for drinking water defluoridation. Capacit. Build. Knowl. Shar. Arm WISA. 2009:1-15. http://www.ewisa.co.za/literature/files/148_101Murutu.pdf

[64] Liang W, Zhan L, Piao L, Rŭssel C. Fluoride removal performance of glass derived hydroxyapatite. Materials Research Bulletin. 2011;**46**:205-209. DOI: 10.1016/j.materresbull.2010.11.015

[65] Yang M, Hashimoto T, Hoshi N, Myoga H. Fluoride removal in a fixed bed packed with granular calcite. Water Research. 1999;**33**:3395-3402

[66] Turner BD, Binning P, Stipp BDS. Fluoride removal by calcite: Evidence for fluorite precipitation and surface adsorption. Environmental Science & Technology. 2005;**29**:9561-9568

[67] Sasaki K, Yoshida M, Ahmmad B, Fukumoto N, Hirajima T. Sorption of fluoride on partially calcined dolomite. Colloids and Surfaces A: Physicochemical and Engineering Aspects. 2013;**435**:56-62. DOI: 10.1016/j.colsurfa.2012.11.039

[68] Patel G, Pal U, Menon S. Removal of fluoride from aqueous solution by CaO nanoparticles. Separation Science and Technology. 2009;**44**:2806-2826

[69] Díaz-Nava C, Olguín MT, Solache-Ríos M. Water defluoridation by mexican heulandite-clinoptilolite. Separation Science and Technology. 2002;**13**:3109-3128

[70] Samatya S, Yüksel Ü, Yüksel M, Kabay N. Removal of fluoride from water by metal ions (Al^{3+}, La^{3+} and ZrO^{2+}) loaded natural zeolite. Separation Science and Technology. 2007;**42**:2033-2047. DOI: 10.1080/01496390701310421

[71] Turner BD, Binning PJ, Sloan SW. A calcite permeable reactive barrier for the remediation of fluoride from spent potliner (SPL) contaminated groundwater. Journal of Contaminant Hydrology. 2008;**95**:110-120. DOI: 10.1016/j.jconhyd.2007.08.002

[72] Turner BD, Binning PJ, Sloan SW. Impact of phosphate on fluoride removal by calcite. Environmental Engineering Science. 2010;**27**:643-650. DOI: 10.1089/ees.2009.0289

Agro-ecology

Chapter 4

Control of Soil pH, Its Ecological and Agronomic Assessment in an Agroecosystem

Danute Karcauskiene, Regina Repsiene,
Dalia Ambrazaitiene, Regina Skuodiene and
Ieva Jokubauskaite

Additional information is available at the end of the chapter

http://dx.doi.org/10.5772/intechopen.75764

Abstract

Lithuania is located in the humid zone, where mean annual precipitation exceeds mean evapotranspiration and soil acidification is an ongoing natural process encouraged by anthropogenic activities. Traditionally, the process may be controlled by different intensity liming. The chapter summarizes the data on long-term liming and fertilization experiments made in Western Lithuania. The object of the investigation is the naturally acid soil, Bathygleyic Dystric Glossic Retisol (texture: moraine loam with clay-sized particles content of 12–14%), and the same soil exposed for more than half a century to different liming and fertilization intensity. Our systematic analysis shows that it is impossible to reach appropriate moraine loam soil conditions for organic matter decomposition, carbon sequestration, soil aggregation, nitrogen fixation, nutrient accumulation, and plant growth by using intensive liming only. It is necessary to co-ordinate proper liming and organic fertilizing. The soil acidity was neutralized (pH_{KCl} 5.9 ± 0.1) and mobile aluminum abolished in the topsoil and subsoil to a 60 cm depth; moreover, the highest amount of soil organic carbon (1.91%), water stable aggregates (59%), intense nitrogen fixation, and highest grain yield was established in the periodically limed (with 1.0 rate $CaCO_3$ every 7 years) soil with 60 t ha^{-1} farmyard manure (FYM) application.

Keywords: soil pH, liming, manuring, microbial activity

1. Introduction

Soil is an integral part of nature. This means that the soil body is not only a recipient but also a donor of its own products and original materials like nitrogen, phosphorus, and carbon to other natural resources—mainly air and water [1, 2]. Modern intensive agriculture based on the use of substances of non-natural origin poses a major threat to nature and primarily to soil. It results in the disturbance of the vital ecological functions of the soil, including chemical buffering capacity, nutrient cycling, biodiversity, organic matter decomposition, as well as esthetic functions. The degradation of soil generally manifests itself by the loss of humus, decrease of biological activity, destruction of structure, increasing compaction, leaching of nutrients, runoff, and acidification [3, 4]. Soil acidification can be accelerated by intensive farming or prevented by sustainable management practices. Soil acidification management includes both neutralization of soil acidity and regulation of the acidification of limed soil. A key aspect in managing acid soils for plant growth is the pH amendment of the soil. Application of lime materials and organic substances could be an effective measure for improving soil chemical properties and the changes of their indexes depend on the amount of liming materials and frequency of application [5, 6]. Spatial patterns of topsoil pH in Lithuania are fundamentally controlled by the soil parent material. Topsoil, subsoil, and parent material collectively influence soil reaction and should not be treated as separate components [7]. Analysis of the pH dynamics in Lithuanian soils over half a century under the influence of hypothetical minimal liming level conducted 50 years ago, suggests that soil acidity neutralization must be an uninterrupted process. Once soil loses its chemical balance, it is less resistant to acidification than initially acidic soil. Liming is a strategic tool for improving acid soils, if the chosen liming intensity allows maintaining soil pH_{KCl} value at a level of 5.7–6.2, which is optimal for many crops. In Lithuania, a long liming tradition has led to a problem of overliming (topsoil pH > 7.0) of moraine loam soil, which results in morphological changes in the soil profile and increased calcium leaching, deterioration of topsoil structure, and decreasing crop yield [8]. This shows that soil pH is a critical parameter that influences the plant-soil-water interfaces. Soil pH impacts on a number of factors affecting microbial activity, like solubility and ionization of inorganic and organic soil solution constituents, and these will in turn affect soil enzyme activity. The soil bioavailability of many nutrients and toxic elements and the physiology of the roots and rhizosphere microorganisms are affected by soil pH [9].

Soil acidity regulates the rate of organic matter mineralization, impacting the release of carbon (C) and nitrogen (N) from soil organic matter. Also, soil acidity has a deleterious effect on the symbiotic relationship between rhizobia and legumes, reducing N fixation and the subsequent supply of N for soil-derived GHG (greenhouse gases) [10]. Changes in the environmental conditions occur in line with alterations in agrophytocenoses. Soil pH influences the changes in the species composition, structure, and plant and environment interaction of the agrophytocenoses.

In summary, the soil genesis and texture, the type of lime materials and their exposure time as well as climatic conditions are very important factors for the liming efficiency on soil pH, structure, and stocks of carbon and nitrogen. There exist opinions that efficiency of agricultural practices on soil pH changes depends on the geochemical environment in which these practices are applied.

In this chapter, we will briefly consider some aspects of the naturally acid moraine loam soil pH management practices and soil-plant interactions at various pH levels under the climatic conditions of West Lithuania.

2. Description of the experiment

2.1. Site

The chapter presents the scientific achievements of half a century research carried out in the Vezaiciai Branch of Lithuanian Research Centre for Agriculture and Forestry. The site of the study was Vezaiciai Branch, Lithuanian Research Centre for Agriculture and Forestry (LAMMC) located in West Lithuania's eastern fringe of the coastal lowland (55°43′N, 21°27′E). The present research was conducted in the crop rotation field devoted to the experiment on long-term liming and fertilization carried out since 1949.

2.2. Soil

Long-lasting geological processes have formed the soils of Lithuania. Prevailing in the country are Luvisols (21%) and Retisols (20.4%). The soil of the experimental site is Bathygleyic Dystric Glossic Retisol (WRB—World Reference Base, 2014) (texture: moraine loam with clay-sized particles content of 12–14%). According to the content of clay particles, the soil profile is differentiated into alluvial and illuvial horizons with diagnostic horizons: Ah-ElB-ElBt-BtEl-BCg. The soil is very acidic (pH_{KCl} 3.9–4.2) in its whole profile up to 160 cm depth and the amount of toxic mobile aluminum is very large, both in the topsoil and subsoil (respectively 100 and 300 mg kg^{-1}), the occurrence of calcareous rock were found up to more than 2 m depth. According to its profile differentiation and all acidity parameters, the soil under study is in priority in terms of the need for liming. The deficiency in clay (<0.002 mm), cations of Ca and Mg and organic colloids is the main factor that influences the low stability of acid topsoil aggregates; hence the soil structure is poor and changeable under various climatic and anthropogenic factors.

2.3. Climate

The climate is moderately warm and humid. Lithuania is characterized by mild winters with frequent thaws, relatively warm springs, moderately warm summers, and warm and wet autumns. The country experiences approximately 70 anticyclones and 100 cyclones annually. Anticyclones predominate in the winter and spring, while cyclones occur in the autumn and summer periods. Lithuania is located in the humid zone, where mean annual precipitation (748 mm) exceeds mean evapotranspiration (512 mm). The mean annual amount of precipitation is more than 800 mm and the average annual air temperature is 6.7 °C in the region of West Lithuania. This region is strongly affected by the maritime climate, because of which it receives the greatest annual amount of precipitation, averaging 923 mm over the last 40 years, compared with the other regions of the country. However, the average annual amount of precipitation has decreased by 7.9% over the past 40 years. The amount of rainfall that falls during the warm period (April-October) is also decreasing (17.9%). The amount of rainfall is particularly important during the spring period when plants start their vegetative/growing season. The variation of precipitation amount in recent years has shown a trend toward reduction in the spring period.

2.4. Experimental design

The investigations were performed at the two stationary liming and fertilizing field trials:

1. The effects of long-term liming on the topsoil properties were estimated using the following experimental design shown in **Table 1**.

2.5. Object of investigation

The naturally acid soil and the same soil exposed to different liming intensity were the objects of the investigation. This long-term field experiment was started in 1949. Applying the long-term liming system (primary, repeated, and periodical liming) in the period 1949–2005 formed the different soil pH levels (**Table 1**). Before liming with 0.5 rate every 7 years during the study period, the soil pH_{KCl} was 5.4–5.9, and in the soil limed with 2.0 rates every 3–4 years, it was 6.4–6.8. The study on the soil structure and organic carbon compounds was conducted in soils that significantly differed in pH.

2.6. Trial history

Pulverized limestone (92.5% of $CaCO_3$) was used for periodical liming on the background of primary and repeated liming with slaked lime. Minimal organic fertilizing (40 t ha^{-1} manure during a seven-course rotation) was undertaken with traditional tillage and intensive crop rotation: sugar beet, spring barley with undersown grasses, perennial grasses (for 2 years), winter wheat, vetch-oats mixture for grain, and pea-barley mixture for forage. In 2008, the long-term experiment was adjusted. In 2005–2013, the soil was not limed. The crop rotation was shortened to four courses: spring barley with undersown grasses, perennial grasses (for 2 years), winter wheat, and spring oilseed rape.

The mineral fertilizing in crop rotation was background ($N_{60}P_{60}K_{60}$), and the fertilizing with organic fertilizers was minimal (40 t ha^{-1} farmyard manure during crop rotation and traditional soil tillage).

Liming intensity	Amount of $CaCO_3$ applied, t ha^{-1}		Total amount of $CaCO_3$ applied, t ha^{-1}
	1949–1998	1998–2005	1949–2005
1. Unlimed (pH_{KCl} 4.0–4.1)	—	—	—
2. Periodical liming using ×0.5 of the liming rate calculated based on the soil hydrolytic acidity (3.3 t ha^{-1} $CaCO_3$) **every 7 years** (pH_{KCl} 5.4–5.9)	18.1	—	18.1
3. Periodical liming using ×1.0 of the liming rate calculated based on the soil hydrolytic acidity (15.0 t ha^{-1} $CaCO_3$) **every 3–4 years** (pH_{KCl} 5.9–6.3)	46.8	7.5	54.3
4. Periodical liming using ×2.0 of the liming rate calculated based on the soil hydrolytic acidity (15.0 t ha^{-1} $CaCO_3$) **every 3–4 years** (pH_{KCl} 6.4–6.8)	89.9	15.0	104.9

Table 1. Experimental design.

2. The effects of long-term liming in combination with organic fertilizers (manure and green fertilizers) on the topsoil properties were estimated using the following experimental design:

Factor A. Soil acidity (pH_{KCl}): (1) unlimed soil (pH_{KCl} pH 4.1–4.3); (2) limed soil (pH_{KCl} 5.8–6.0).

Factor B. Organic fertilizers: (1) without organic fertilizers (control treatment); (2) green manure or plant residues; (3) farmyard manure 40 t ha^{-1}; (4) green manure (on a background of 40 t ha^{-1} farmyard manure (bkgd of FYM 40)); (5) farmyard manure 60 t ha^{-1}; and (6) green manure (on a background of 60 t ha^{-1} farmyard manure (bkgd of FYM 60)).

In a long-term experimental trial of farmyard manure rates starting from 1959 to 2005, 80 and 120 t ha^{-1} of farmyard manure were incorporated in two applications divided into equal parts for the seven-course crop rotation (for winter wheat and fodder beet). After the reconstruction of the trial in 2005, 40 and 60 t ha^{-1} of farmyard manure were incorporated in a single application (for winter wheat) in the five-course crop whereas in the fourth and sixth treatments, manure was not applied. Solid cattle manure was used, containing 14.53% of dry matter, 17.83% of organic matter, 0.42% of total nitrogen N, 0.27% P_2O_5, 0.67% K_2O, 2668 mg kg^{-1} Ca, 692 mg kg^{-1} Mg, pH_{KCl} 8.5.

The following alternative organic fertilizers were employed: in 2010, the aftermath of swards was disked in at 15 cm and plowed in at 20 cm depth. In 2011, after wheat harvesting, the straw was chopped, incorporated at 15 cm and plowed in at 20 cm depth. In 2012, the green mass of lupine and oats was disked in at 15 cm and plowed in at 20 cm depth after lupine pods had reached milk maturity. In 2013, after rape harvesting, the stubble and straw were chopped and incorporated by a cultivator at 15 cm and plowed in at 20 cm depth.

On limed background, in 2010, liming was applied repeatedly using pulverized limestone at one rate according to the hydrolytic acidity. All treatments were equally fertilized with mineral fertilizers (background fertilization). Fertilizer $N_{60}P_{60}K_{60}$ have been applied for winter wheat and spring barley stands, $N_{30}P_{60}K_{60}$ for lupine-oats mixture, and $N_{60}P_{90}K_{120}$ for winter rape stands. Fungicides and insecticides were used in case of necessity; herbicides were not used at all. Conventional soil tillage was applied. The acidic soil had been periodically limed and manured for a period of 47 years. During the period of the study, the soil received: 38.7–36.5 t ha^{-1} $CaCO_3$; 840 t ha^{-1} cattle manure, 2740 kg ha^{-1} mineral nitrogen; 3030 kg ha^{-1} phosphorus, 3810 kg ha^{-1} potassium.

3. Management of soil acidity

Acidification of soil is a continuous naturally occurring process in soil formation. It is promoted by natural and anthropogenic factors. Soil acidification management involves both neutralization of soil acidity and regulation of the acidification of limed soil. The pH level in agricultural soils is strongly influenced by the fertilizers and pesticides applied, crops grown and their sequence in a crop rotation, as well as by tillage intensity. Liming is the most efficient measure used to neutralize soil acidity. Liming materials alter the mobility of some biogenic elements and their buildup in the soil. Application of lime materials and organic substances could be an effective measure to improve the chemical properties of moraine loam soil, but the changes in their values depend on the amount of liming materials and frequency of application. Studies conducted in Lithuania have shown that the effect of initial liming lasts for over 30 years. The published data indicate that the change in soil acidity indicators after liming primarily depends on whether the liming is performed for the first time or repeatedly. Within a period of 50 years after primary liming, the pH values return to the initial ones and practically do not differ from those of acid soils; however, it was noticed that the soil limed once even with the lowest 0.5 rate (according to hydrolytic acidity) tended to acidify more rapidly compared with the soil that had never been limed. This suggests that once the chemical balance of the soil has been disturbed, it becomes less resistant to environmental impacts than a naturally acid soil. Therefore, from the soil conservation viewpoint, liming should be an uninterrupted and systematic process in the agroecosystem. The results of such systematic (primary, repeated, and periodical) liming are significantly changed pH values in the upper and deeper soil horizons and alterations in the mobile Al contents (**Figures 1** and **2**).

Regular liming (during a 35-year period) resulted in a decrease in the hydrolytic acidity and mobile aluminum in a loam soil and an increase in the pH in the soil profile up to 100 cm depth. Periodical liming every 7 years at a rate of 0.5 by hydrolytic soil acidity (3.8 t ha^{-1} CaCO$_3$) allowed maintaining the soil at a medium acidity level (pH$_{KCl}$ 4.8–5.1); when the soil was intensively limed at a rate of 1.0 every 3–4 years, mobile aluminum was abolished and pH$_{KCl}$ in the upper layers of the soil reached close to neutral, 6.4–6.7. Long-term regular soil liming changes the acidity not only in the topsoil, but, because of the migration of calcium and magnesium cations, the chemical properties in the subsoil horizons change as well. When liming is intensified to 2.0 rates every 3–4 years, the pH$_{KCl}$ value in the topsoil reaches 6.9–7.2, i.e., the acid soil becomes neutral and acidification is hindered in the E1B horizon up to 40 cm depth (**Figure 2**).

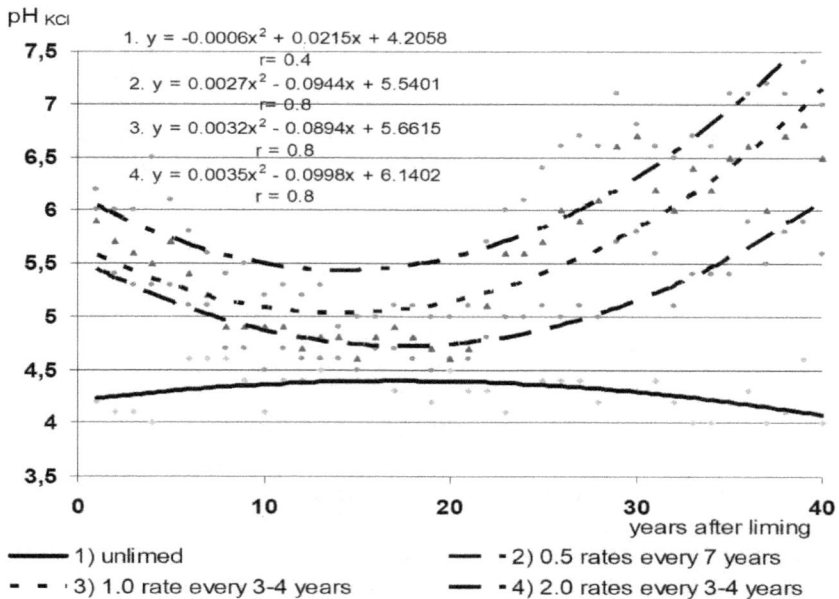

Figure 1. Long-term liming effect on topsoil pH changes.

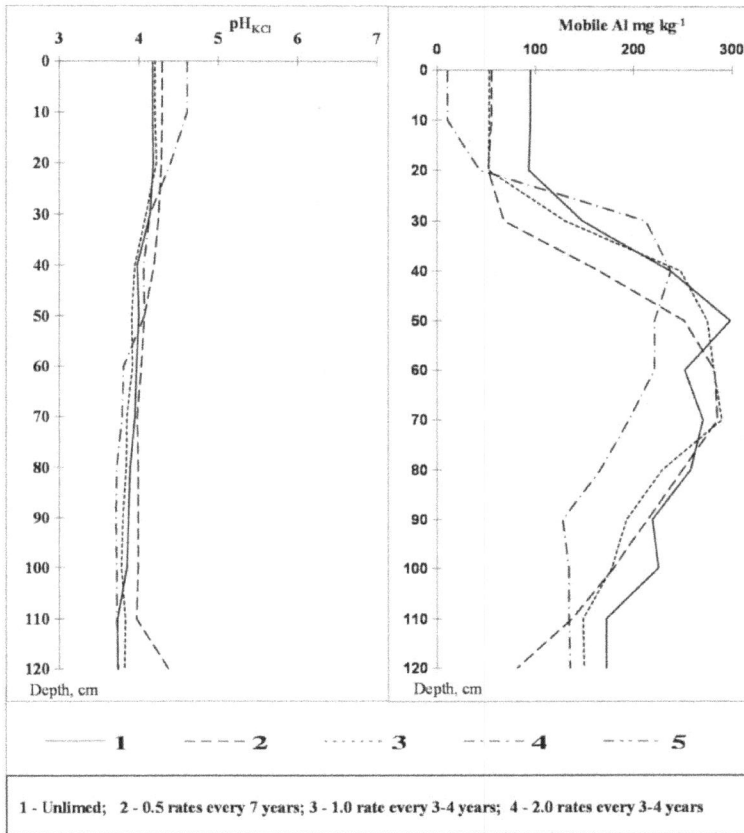

Figure 2. Long-term liming effect on pH and mobile Al changes in soil profile.

Long-term intensive liming causes significant morphological changes in the soil profile, and increases the area of podzolic veins in the profile wall. Ozeraitiene has documented that the highest leaching of calcium occurs and soil structure deteriorates in intensively limed soil [8]. This shows that it is impossible to reach optimal soil conditions for plant growth by using intensive liming only. It is necessary to combine liming with organic fertilizing. A number of different mechanisms have been proposed to explain the positive effect of organic residues, and their decomposition products in raising soil pH and/or complexing phytotoxic Al and thus improving the fertility of acid soil [11–14].

According to Mokolobate and Haynes, the addition of organic residues to acid soils is potentially a practicable low-input strategy for increasing soil pH, decreasing concentration of phytotoxic Al, and reducing lime requirements [14]. This statement was based on our data obtained in naturally acid soil with incorporation of farmyard manure. A systematic fertilization of naturally very acidic soils with farmyard manure at a 60 t ha^{-1} rate every 3–4 years for over five decades resulted in decrease in mobile Al by 2.3 times, i.e., from 117.0 to 50.5 mg kg^{-1} and in an increase in the pH in the topsoil by 0.2 units (**Figure 3**).

The soil acidity neutralizing effect of solid manure showed up only after 20 years of its application, but its significant effect manifested itself the next year after application, when pH_{KCl} values reached 4.8–5.3 and mobile Al content decreased to <50 mg kg^{-1} (**Table 2**).

A combination of farmyard manure application (60 t ha^{-1} every 3–4 years) with periodical liming (1.0 rate of $CaCO_3$ according to hydrolytic soil acidity every 5–7 years) stabilizes the acidification process (a pH of 5.6–7.1 is maintained) in the arable layer. The findings of the long-term experiments indicate that liming alone was less efficient for soil acidity indicators than its combination with farmyard manure. The obtained results substantiate the research evidence reported by Teit that liming materials and farmyard manure are agronomic practices that cannot replace each other but can complement each other [15].

Liming materials and their combinations with farmyard manure (FYM) reduce the amount of mobile aluminum to a level that is non-toxic to plants (1.17–1.70 mg kg^{-1}). An analysis of the long-term data revealed that the amount of mobile Al after liming and FYM fertilizing decreased to 0.0–0.3 mg kg^{-1}, while after 6–7 years, an increase to 1.3–6.0 mg kg^{-1} was determined.

The neutralization of soil acidification in the topsoil and subsoil up to 60 cm depth was achieved by periodical liming with 1.0 rate every 7 years in combination with the application

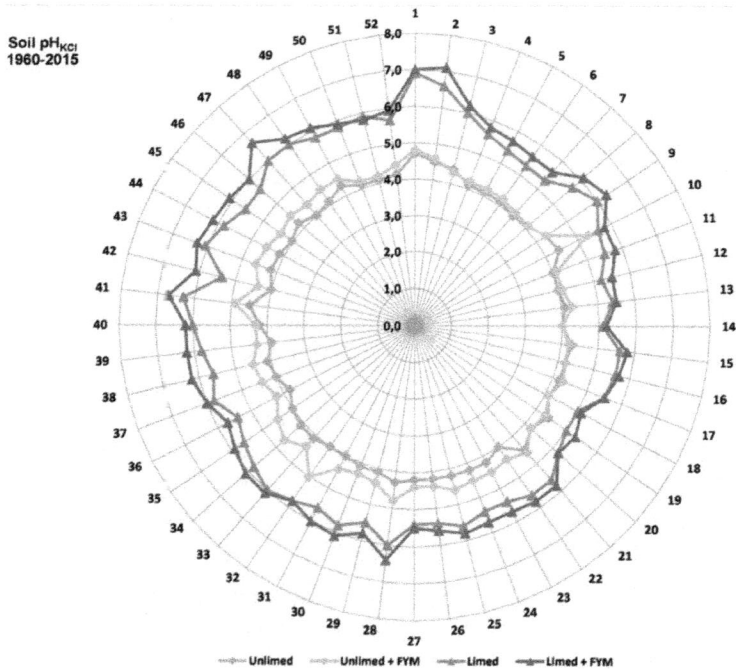

Figure 3. Long-term liming and manuring effect on topsoil pH changes.

Treatment.	Arithmetic mean $\bar{x}$	Standard error of the mean $S\bar{x}$	Minimal value of indicators Min.	Maximal value of indicators Max.	Standard deviation SD	Coefficient of variation $V\%$
pH (units)						
Unlimed	4.16	0.02	3.80	4.70	0.17	4.06
Unlimed + FYM	4.40	0.03	4.00	5.30	0.25	5.62
Limed	5.66	0.05	5.00	6.90	0.36	6.34
Limed + FYM	5.93	0.06	5.10	7.10	0.41	6.87
Mobile Al (mg kg^{-1})						
Unlimed	117.0	2.84	63.90	169.7	20.28	17.34
Unlimed + FYM	50.5	4.33	13.10	107.5	30.90	61.20
Limed	1.70	0.22	0.00	7.60	1.55	91.41
Limed + FYM	1.17	0.18	0.00	6.0	1.25	107.1

Table 2. Results of statistical analysis of agrochemical (pH and mobile Al) indicators.

of 60 t ha^{-1} manure every 3–4 years in the seven-course crop rotation system (**Figure 4**). The effects of long-term (1959–2005) liming in combination with cattle manure application on soil pH and mobile aluminum were investigated in the whole soil profile up to a 100 cm depth. Acid soil had been periodically limed and manured at different intensities for 47 years. During the entire period of the study, the soil received 38.7–36.5 t ha^{-1} CaCO$_3$; 840 t ha^{-1} cattle manure; 2740 kg ha^{-1} mineral nitrogen; 3030 kg ha^{-1} phosphorus; 3810 kg ha^{-1} potassium. The findings suggest that long-term (47 years) periodic liming in combination with the application of cattle manure significantly altered the chemical properties within the entire soil profile.

In the topsoil and subsoil, up to 60 cm depth, the soil acidity was neutralized by systematic liming at 1.0 rate every 7 years in combination with the application of 60 t ha^{-1} manure every 3–4 years. Long-term periodical liming in combination with manuring improved the chemical properties of acid soil profile in the ElB and ElBt horizons.

The rate of acid soil neutralization depends on the particle size of the liming materials. According to the size, liming materials can be classified as powdered, granulated, pelletized, etc. Powdered liming materials show the most rapid action; however, their application is rather complicated because special machinery is required in order to spread them evenly. Granulated liming materials react longer with soil and the duration of their action is longer than that of powdered ones [16]. In Lithuania, acid soils for more than three decades (1964–1994) were systematically limed every 5–7 years with conventional liming material— powdered limestone [17].

Recently, granulated liming materials have been increasingly used in European, including Lithuania, and other countries for maintenance liming of soils [18, 19].

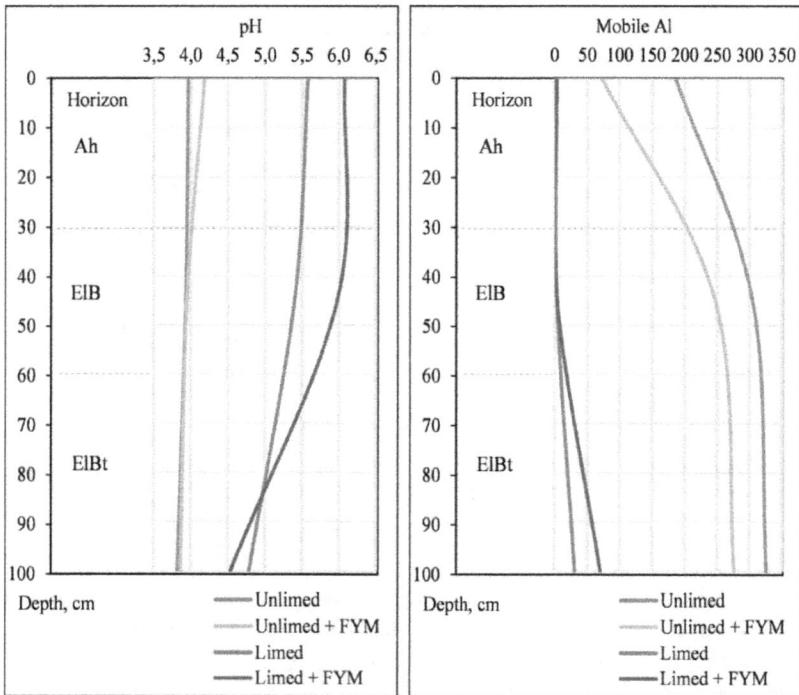

Figure 4. Long-term liming and manuring effect on pH and mobile Al changes in soil profile.

4. Soil pH and microbial prevalence

The provision of energy to the microbial community by root exudates, dead roots, and intense microbial metabolic activity benefits mineralization capacity, improved soil structure, and plant growth conditions. Mineral nutrients, organic matter (including fresh plant material), and growth factors exist within appropriate temperature, moisture, redox potential and pH ranges for microbial growth [20–22].

The majority of soil systems tend to fall in the soil pH range fall below 5.5 in the agroecosystem. Yet, acidic soils constitute a major portion of the world soil resources. West Lithuania also falls in the region of acidic soils. Acid conditions present a stressful environment not only for plants but also for the microbial community. In West Lithuania, albeluvisol (Bathygleyic Dystric Glossic Retisol (WRB 2014) has been investigated for nearly six decades. Soil liming and nutrition with various organic fertilizers: incorporation of plants residues (treatments 2, 4, 6) and 40 t ha^{-1} (treatments 3, 4), or 60 t ha^{-1} (treatments 5, 6) of farmyard manure created different conditions for the functioning of soil microorganisms and carbon sequestration in the soil. The experimental findings suggest that a significant increase in the pH value was achieved due to liming, and the concentration of C significantly increased when farmyard manure together with the incorporation of plant residues had been used for crop nutrition. During the experimental period, soil chemical indicators changed and microorganism communities, which determined the biochemical processes in the soil, formed. The following processes were analyzed: greenhouse gas release—CO_2 emission from the soil—and groups of microorganisms participating in the organic matter decomposition processes and being able to assimilate mineral nitrogen [23]. The CO_2 emission from the soil is also an indicator of the general biological activity of soil microorganisms. Our study showed that with a significant reduction in soil acidity resulting from liming, a significant increase in CO_2 emission occurred. An increase in the pH value (factor A) by an average 1.62 units significantly increased the content of ammonifying, mineral nitrogen assimilating, and spore forming bacteria and was essential in reducing the number of fungi, because more compounds are formed in the soil, which are available to bacterial microflora (**Table 3**). According to the averaged results, increasing concentration of carbon (factor B) in the soil enhanced CO_2 emission; however, a more detailed analysis revealed that in acid soil, CO_2 emission increased in all organic fertilization cases, but when soil acidity decreased, organic fertilization was no longer an effective means. As a result of uneven fertilization, higher concentration of carbon accumulated in the soil and a significant increase was observed in the abundance of ammonifying and mineral nitrogen assimilating microorganism populations, indicating more favorable conditions for the functioning of the microbial community in the soil. The same regularity was established for sporogenes bacteria, which are also very active in the organic matter mineralization processes. However, the abundance of fungi populations was substantially higher only with the integrated incorporation of manure and plant residues in the soil.

Such complex fertilization introduces a larger amount of more complex compounds into the soil, which can be mineralized by the complex enzyme system of fungi.

Symbiotic biological nitrogen fixation is one of the processes that describe the ecological balance of soils. Acid soils are characterized by poor diversity and activity of microorganisms [24]. None or insignificant amounts of *Azotobacter chroococcum* and *Trichoderma lignorum* are found. Nitrification process, which ensures the availability of nitrogen compounds to plants and other microorganisms, reaches the highest intensity when soil acidity was diminished to pH 6.7. Long-term studies indicate that the optimal pH indicator value for various physiological groups of microorganisms is different: for ammonifying and sporogenes bacteria, it is 6.2; for streptomycetes bacteria, 6.7; and for mineral nitrogen assimilating bacteria and micromycetes, it is 6.0 [25].

Soil reaction is one of the most important factors influencing legumes and *Rhizobium* symbiosis. Greater concentration of H^+ ion increases the solubility of Al, Mn, and Fe; these elements may become toxic to plants. However, it is known that *Rhizobium leguminosarum bv. trifolii* can survive in very acidic soils due to low acidity microzone formation; hence, rhizobia can be isolated from these soils [26]. All living organisms require nitrogen to make proteins, enzymes, and other cellular components. Seventy-eight percent of the atmosphere contains nitrogen that is required by all living organisms, but the gaseous form is unavailable for most organisms. Those that can take up nitrogen gas, do so through a process of nitrogen

Treatment	Soil pH and carbon (%) variation in soil (mean indices)	CO_2 emission, mg g^{-1} abs. dry soil day^{-1}	Ammoni-fying bacteria, CFU × 10^6 g^{-1} abs. dry soil	Nitrogen assimila-ting bacteria, CFU × 10^6 g^{-1} abs. dry soil	Micromy-cetes, CFU × 10^4 g^{-1} abs. dry soil	Sporoge-nes bacteria, CFU × 10^4 g^{-1} abs. dry soil
Soil pH, liming (factor A)						
Unlimed soil (pH_{KCl} 4.1–4.3)	4.24–4.53 (4.35)	0.0271	5.9	4.6	4.3	3.1
Limed soil (pH_{KCl} 5.8–6.0)	5.87–6.15 (5.97**)	0.0288*	7.9*	7.7*	2.7*	7.6*
$LSD_{0.05}$		0.00095	0.47	0.268	0.107	0.23
C_{org}, organic fertilizers (factor B)						
Without organic fertilizers	1.328–1.459 (1.41)	0.0272	5.8	4.8	3.5	3.9
Green manure or plant residues	1.408–1.528 (1.47)	0.0273	6.8	4.8	3.5	3.8
Farmyard manure 40 t ha^{-1}	1.475–1.657 (1.58**)	0.0298	7.1*	6.8*	3.5	6.4*
Green manure (bkgd of FM 40)	1.493–1.718 (1.62**)	0.0273	7.1*	6.9*	3.2	5.4*
Farmyard manure 60 t ha^{-1}	1.478–1.649 (1.59**)	0.0277	7.4*	6.9*	3.6	6.8*
Green manure (bkgd of FM 60)	1.382–1.540 (1.48)	0.0284	7.2*	6.8*	4.1*	6.1*
$LSD_{0.05}$		0.00213	1.057	0.599	0.239	0.514

*Significant at $P \leq 0.05$.
**Significant at $P \leq 0.01$.
CFU, colony forming unit.

Table 3. The effect of different soil pH (liming) and supply of carbon (organic fertilization) on microbiological indicators.

fixation. Many species can fix carbon though photosynthesis (all green plants, algae, some bacteria) but only few organisms can fix nitrogen. Nitrogen fixing organisms can be free-living or symbiotic. Symbiotic associations developed after the evolution of terrestrial green plants. Free-living nitrogen fixers are among the most ancient organisms. Symbiotic associations between species of nitrogen-fixing bacteria and green plants result in considerable amounts of nitrogen being fixed. Among the best studied of these symbiotic associations are those among various species of *Rhizobium* and other species of legumes. These associations are very specific. Phylogenetic analysis of the rhizobia based on the 16S rRNA subunit places the species into three genera: *Rhizobium*, *Bradyrhizobium*, and *Azorhizobium*. Metagenomic analysis of differently fertilized and tilled Bathygleyic Distric Glossic Retisols suggest that Rhizobium accounts for 0.3–0.4%, i.e., 39,358–64,767 operation taxonomy units (OUT) of the 131,194–161,917 identified OUT. Research on symbiotic nitrogen fixation has been carried out for 50 years in West Lithuania. Experimental results obtained by various researchers were comprehensively presented in various scientific publications and summarized by Lapinskas [27].

Symbiotic nitrogen fixation depends on the occurrence and survival of rhizobia in the soil and also on their efficiency. Soil reaction is one of the most important factors that influence legume and *Rhizobium* symbiosis. The data of long-term research evidenced that additional inoculation with rhizobia bacteria resulted in 6–16% increase in dry matter yield, and for some species, for example, goat's rue and soy, the dry matter yield increase amounted to 119–165% [28]. Such differences are primarily determined by the occurrence of the specific rhizobia and the activity of their enzymes in the soil. This is especially relevant in soils, which, due to the intensive use of mineral fertilizers, acidify, thus contributing to the survival and proliferation of atmospheric nitrogen fixing bacteria. *Sinorhizobium meliloti* and *Rhizobium galegae* bacteria, which form symbiosis with certain plants, for example, lucerne and Caucasian goat's rue, are highly sensitive to acid soil and soluble Al. The acidification inhibits the root-hair infection process and nodulation. The prevalence of the main species of rhizobia (*Rhizobium leguminosarum bv. trifolii* and *Rhizobium bv. viciae, Sinorhizobium meliloti* and *Rhizobium galegae*) was established in 400 different soil samples in Lithuania [28]. A dilution method was used for legume inoculation in sterile conditions (**Figure 5**). Estimation of the rhizobia population abundancy suggested that the range of the pH value is not the same for different bacteria: in very acid soil (pH 4.1–4.5), *Sinorhizobium meliloti* was not detected, while *Rhizobium galegae* bacteria were detected only when the soil pH value ranged from 5.1 to 5.5. Such low pH was best tolerated by *Rh. leg. bv. trifolii* (14.2 × 10^3) and *Rh leg. bv. viciae*, but even their populations were not rich. The optimal pH range for *Rh. leg. bv trifolii* was 5.6–7.0, for *Rh leg. bv viciae* 6.1–7.0. *Sinorhizobium meliloti* was the most numerous at a pH range of 6.6–7.0 and only *Rhizobium galegae* were most abundant at pH > 7.0. It is known that the symbiotic efficiency of different legumes and *Rhizobium* bacteria strains depends on the environmental conditions, mineral nitrogen concentration, and soil pH changes, which essentially determine the activity of nitrogenase enzyme [29]. The activity of nitrogenase was best investigated in the red clover association (**Figure 6**). The gas chromatography assay, used to determine the activity of nitrogenase enzyme, responsible for the biological nitrogen fixation in the red clover and *Rh. leg. bv. trifolii* association suggested that in an acid soil, the change of the pH value from <4.75 to 6.0 was insignificant for nitrogenase activity, but an increase in the pH value above 6.25 resulted in a nearly 3-fold increase in nitrogenase activity. During the summer period, in the same soil, even at a soil pH < 4.75, nitrogenase fixed biological nitrogen 10 times more intensively compared with the autumn period. As the pH value increased to 5.25, nitrogenase

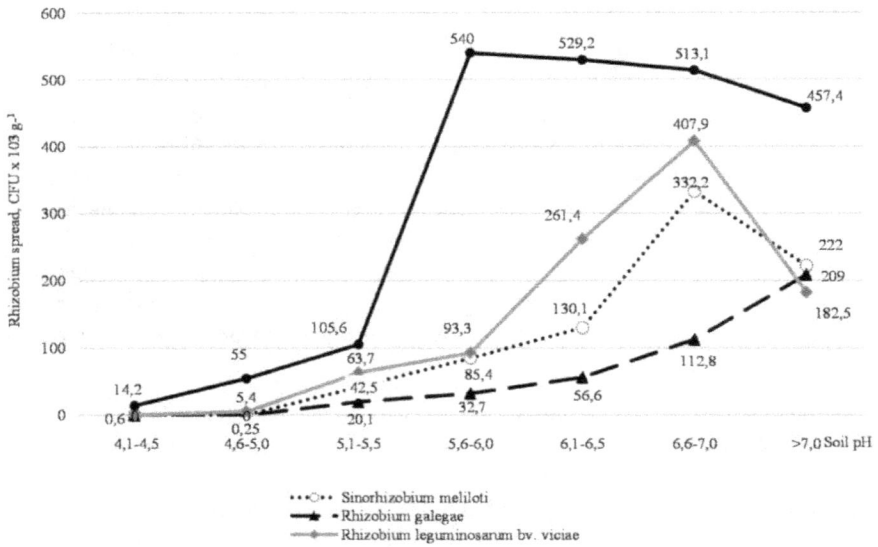

Figure 5. The impact of the soil pH on *Rhizobium* spread (×10³ CFU g⁻¹).

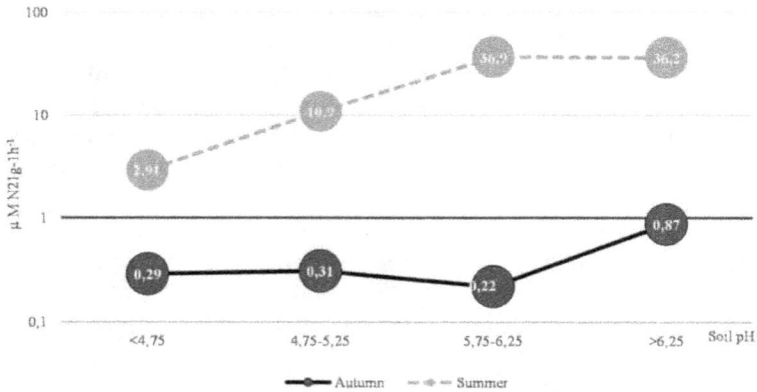

Figure 6. The activity of nitrogenase as influenced by soil pH and season.

activity was 10.9 µ M N₂1 g⁻¹ h⁻¹ and was 35 times higher than in autumn. With the change of the pH value from 5.75 to >6.25, nitrogenase activity increased more than 40-fold. It was noted that soil acidity was a more stressful factor for nitrogenase activity in autumn than in summer.

This determination of the tolerance of different rhizobia species for low pH is an opportunity to use biological preparations more efficiently and to select plants that are resistant to soil with a lower pH [27]. Biological fertilizers have recently become a very popular product used to improve the vital functions of soils. This is especially relevant in ecologically sensitive agro-climatic regions, like West Lithuania, which receive a lot of rainfall and where naturally acid soils predominate. Application of various biological measures, including additional incorporation of humic and amino acids, trace nutrients, and various species of microorganisms that supplement the complex of soil microorganisms, is a common practice [29, 30].

5. Soil pH effect on organic carbon and water stable aggregate content

Soil organic matter is the chief indicator of soil quality and ecological stability. Consequently, carbon accumulation in stable forms not only supports and enhances the organic matter content in the soil, but also exerts a positive impact on the soil quality and the entire ecosystem [31]. Taking into account the environmental conditions and land use, soil can be the source of both carbon accumulation and release. The directions of these processes depend on the intensity of the anthropogenic load present in the specific natural conditions. Soil liming and organic fertilization are one of the most common ways to improve carbon sequestration in the soil [32, 33]. Negative statistically significant effect of periodical liming at 0.5 and 2.0 rates on SOC (soil organic carbon) content was determined (**Figure 7**). In the unlimed treatment, the SOC content was 1.45%, while in the periodically limed treatment at 2.0 rates every 3–4 years, it was by 0.18% lower. A positive statistically significant effect of fertilization on soil organic carbon content in the soil was established. The content of SOC was 1.47% in the non-fertilized treatment and in the fertilized treatments, it varied from 1.59 to 1.91%. The highest amount of soil organic carbon (1.91%) was obtained in the limed soil applied with FYM. A similar content of SOC was determined after incorporation of alternative organic fertilizers (wheat straw, oilseed rape residues, roots, stubble, and perennial grasses). However, in contrast to incorporated farmyard manure, the largest part of alternative organic fertilizers is composed of nitrogen compounds that are rapidly mineralized in the soil.

Decomposition of FYM includes longer phases of mineralization and nitrogen immobilization, thus increasing the amount of stable organic compounds and accumulation of humus in comparison with the use of plant residues. SOC and dissolved organic carbon (DOC) are important indices for soil organic matter (SOM). Dissolved organic carbon (DOC) is a key component of the active SOM pool and serves as a source and sink of soil nutrients and/ or as an ecological marker to understand soil fertility. Soil liming with or without fertilization could affect the content of DOC, leading to alterations in the formation of complexes between organic ligands and metals, and SOM sequestration. On the other hand, increased decomposition of stabilized material induced by addition of fresh organic material triggering microbial activity can result in higher C losses from soil. These indices are influenced by agricultural practices [34, 35]. The content of DOC is closely related to soil- and management-associated factors. Liming at different intensities did not show any statistically significant effect on the accumulation of humic acids in the soil. However, intensive soil liming (at 2.0 liming rate) tended to increase the content of humic acids (**Figure 8**). Fulvic acids predominated in unlimed soil. This treatment was found to have the highest amount of fulvic acids (0.562% of C). It could be associated with a slow humification processes in the soil, due to the

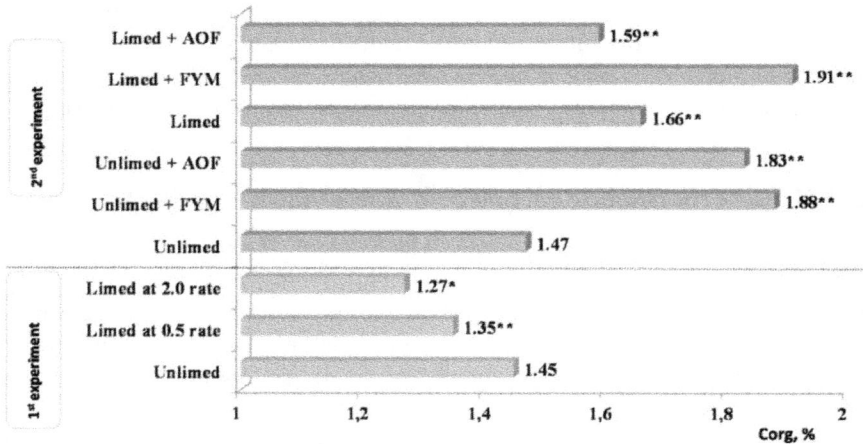

Figure 7. Effect of soil liming and fertilization on the amount of organic carbon in the topsoil, 2011–2013 average data. *Note:* * and ** Significantly different from control (unlimed) ($P < 0.005$) and ($P < 0.001$); FYM, farmyard manure; AOF, alternative organic fertilizers.

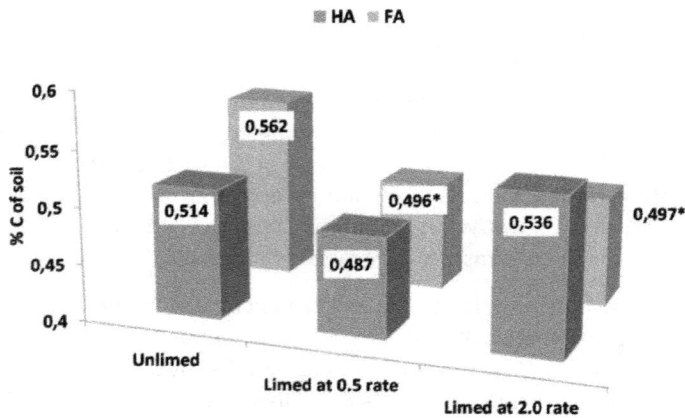

Figure 8. Liming effect on the total content of humic and fulvic acids in soil, % C (where HR, humic acids; FR, fulvic acids). *Note:* * Significantly different from control ($P < 0.005$).

low carbon content, when the content of fulvic acids is always considerably higher compared with humic acids in the soil. Soil liming significantly decreased the amount of fulvic acids. Soil amendment with organic fertilizers significantly increased the amount of humic acids, which indicates an increase in organic matter fixation and its stability in the soil. Seeking to assess the relationship between labile and stable humic acid amount and soil pH, we conducted a correlation and regression analysis. Mobile humic acid fraction (HR1) was assigned as labile humic acids, while humic acids fractions bound with calcium (HR2) and soil minerals (HR3) were stable fractions. It was determined that increasing soil pH decreased labile and increased stable humic acid amounts in the soil (**Figure 9**).

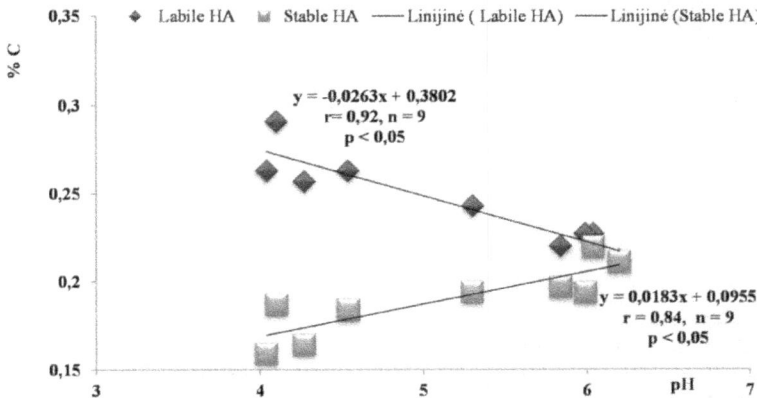

Figure 9. The relationship between labile and stable humic acids amount and soil pH.

Soil aggregates stability is a crucial soil property affecting soil sustainability. Physical protection of soil organic matter by aggregates is considered to be an important mechanism for soil carbon stabilization. Due to the small amount of water-stable aggregates, the structure of acid moraine loam soil is poor and depends on various climatic and anthropogenic factors. The findings of our study evidenced that long-term (for 56 years) systematic liming by 0.5 rate every 7 years and 2.0 rates every 3–4 years on the background of minimal organic fertilizing and traditional tillage affected the structure of moraine loam soil (**Table 4**). According to the data obtained over a period of 11 years, the largest amount (68.4–72.1%) of water stable aggregates (>0.25 mm) was present in the intensively limed soil under perennial grasses and winter wheat in 2002–2003.

A 28–34% increase in the content of water stable aggregates was recorded in limed soil compared with unlimed soil. These data suggest that perennial grasses grown in limed soil can effectively improve the structure of moraine loam soil. During the study period, a decrease in water stable aggregates was determined in both acid (from 57 to 25%) and limed (from 72 to 29%) soil. These findings indicate that the climatic conditions (swelling of clay under warm and wet conditions in autumn–winter periods and drying and rewetting cycles in spring-summer periods) have a significant negative impact on the formation of the aggregates. The negative effect of these weather conditions on soil aggregates are observed by other researches [36]. Also, the decrease of water stable aggregates in moraine loam soil is caused by the intensive anthropogenic activity inducing a decrease of organic carbon. Any agricultural activity which increases organic matter content in the soil has a direct effect on the increase of soil aggregate stability. The data of the soil structure in the topsoil and subsoil showed that long-term manuring had a positive effect on water stability of soil aggregates (**Figure 10**).

Under the influence of long-term application of farmyard manure, the content of water stable aggregates >0.25 mm increased by 48.1% and amounted to 52.3%, compared with the treatment with no manure. An 81.6% increase in the content of aggregates was determined in soil that had been fertilized with manure for a long time and optimal pH had been maintained by periodic liming.

Treatment	Investigation year and vegetation type										
	1996	1997 V + O	1998	2002 G	2003	2004	2005 V + O	2008	2009	2010	2011
	W		P + B		W	P + B		B	G	T	R
Unlimed	57.3	50.4	56.5	56.2	50.9	40.1	42.0	34.2	37.7	26.0	24.8
0.5 rate every 7 years	51.5	48.9	53.4	64.7	63.8*	53.5*	51.4*	47.1*	43.8	32.0	29.8
2.0 rate every 3–4 years	56.0	55.9*	58.4	72.1*	68.4*	50.5*	50.6	42.4	53.2*	30.5	28.8

*Significantly different from control ($P < 0.005$).

W, winter wheat; V + O, mixture of vetch and oat; P + B, mixture of peas and barley; G, perennial grasses; B, spring barley; T, winter triticale; R, spring rape.

Table 4. The effect of periodical liming on the content (%) of water stable aggregates (>0.25 mm).

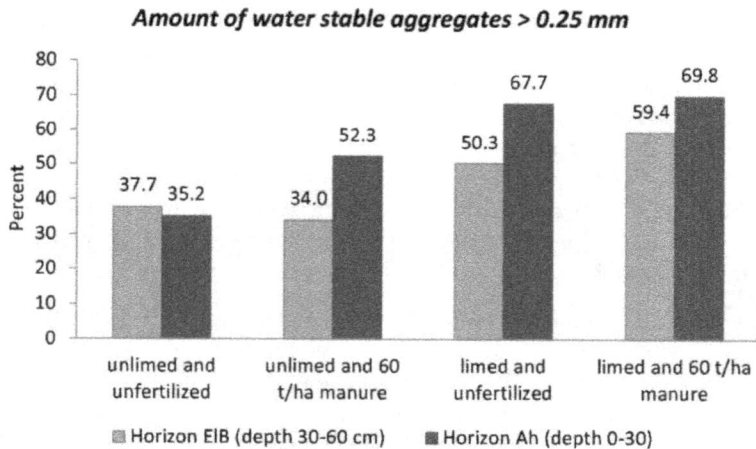

Figure 10. Effects of long-term liming, manuring, and their combination on the change of water stable aggregates in the soil profile.

The content of water stable aggregates both in the naturally acid and in limed soil was 30.3% higher in the upper Ah horizon compared with the deeper ElBt (30–60 cm) layer. This is related to the impact of plant roots present in the upper topsoil layer and higher organic matter content, which are relevant factors for the formation of water stable aggregates. Liming and its combination with the application of manure had a positive influence on the formation of water stable aggregates in the deeper 30–60 cm soil layer as well. In the limed and the limed and manure-applied ElBt soil horizon, the content of water stable aggregates was 33.5 and 57.2% respectively higher than in the unlimed and unfertilized soil. This was associated with a slight increase in calcium ions and organic carbon resulting from the long-term application of crop and soil management practices. The highest content of water stable aggregates (59.4%) was established in the limed and manure-applied soil. Manure application alone did not have any impact on the formation of water stable aggregates in this soil layer.

6. Soil pH effect on crop productivity and weediness

Liming of acid soils, particularly periodical liming and FYM fertilization, is the most important amelioration means of Dystric Albeluvisols, which significantly changes their ecological state. Long-term research enables assessment of the effects of various agronomic practices on the productivity of various agricultural crops. Soil acidity limits the yield of many agricultural crops. A low amount of base cations, particularly toxicity of calcium and aluminum, affect root growth and plant water and nutrient uptake; crop yields most often decline due to acid soils [37]. In the field experiment, winter wheat was grown after perennial grasses. According to the experimental design, before wheat sowing, individual plots were fertilized with solid cattle manure at rates 40 t ha^{-1} and 60 t ha^{-1}. The phytometric indicators that form wheat productivity (ear length and number of grains per ear and 1000 grain weight) were significantly higher in the limed soil and this had a direct effect on grain yield (**Table 5**). In limed soil, winter wheat grain yield was 1.8 times higher and that of straw 1.3 times higher than in unlimed soil. The highest winter wheat grain yield, 3.45 t ha^{-1}, was produced in soil fertilized with 60 t ha^{-1} FYM.

Lupine performs best in acid soils (pH 4.5) and worst in neutral or alkaline soils. It is susceptible to too high concentration of calcium ions, particularly at the beginning of growth. Oats are moderately susceptible to soil acidity. They perform best at soil pH 5.5–6.0 and satisfactorily at pH 4.5. Lupine was grown after winter wheat. Both in unlimed and limed soil, the total number of emerged plants were very similar 156–157 ha^{-1}. The highest dry matter yields of the lupine-oats mixture (9.60-10.0 t ha^{-1}) were recorded in soil fertilized with 60 t ha^{-1} FYM and green manure (bkgd of FYM 40).

Like winter wheat, spring barley is also very susceptible to soil acidity, particularly to mobile aluminum. In limed soil, the values of yield-forming indicators (number of productive and unproductive stems, number of grains per ear, and 1000 grain weight) were significantly higher than in unlimed soil. In limed soil, barley grain yield was 2.2 times higher than in acid soil. The highest yields (3.93–4.00 t ha^{-1}) were produced in soil fertilized with 60 t ha^{-1} FYM and green manure (bkgd of FYM 40). Similar trends were identified for straw yield.

Research showed that in a naturally acid soil (pH 4.1–4.3), in the crop of winter wheat before harvesting, the number of weeds was higher by 79.8%, in lupine–oats mixture by 35.7%, in spring barley crop by 29.1% compared with limed soil (pH 5.8–6.0) (**Table 6**). Similar trends were established when analyzing the data of dry matter mass of weeds. Irrespective of the soil pH, the effect of different organic fertilizers manifested itself best for the second member of the crop rotation (lupine-oats mixture). The highest number of weeds (356.7 m^{-2}) and weed mass (122.7 g m^{-2}) were established in the treatment where the mixture was grown without organic fertilizers. The number and mass of weeds were on average 1.3 and 1.9 times lower in the

Treatment.	Winter wheat yield t ha⁻¹		Lupine-oats mixture DM, t ha⁻¹	Spring barley, t ha⁻¹	
	Grain 14% moisture	Straw DM		Grain 14% moisture	Straw DM
Soil pH, liming (factor A)					
Unlimed soil (pH_{KCl} 4.1–4.3)	1.90	2.57	8.82	1.99	1.72
Limed soil (pH_{KCl} 5.8–6.0)	3.51**	3.26**	9.25	4.34**	3.00**
Organic fertilizers (factor B)					
Without organic fertilizers	1.83	2.04	8.55	2.17	1.41
Green manure or plant residues	1.91	2.04	8.55	2.31	1.72
Farmyard manure 40 t ha⁻¹	3.04**	3.34**	8.81	3.62**	2.78**
Green manure (bkgd of FM 40)	3.16**	3.44**	10.0**	4.00**	2.52**
Farmyard manure 60 t ha⁻¹	3.45**	3.36**	8.70	3.93**	2.96**
Green manure (bkgd of FM 60)	2.86**	3.29**	9.60	3.00**	2.76**
Interaction of factors A × B					
	**	**	ns	**	**

ns, not significant; DM, dry matter.*Significant at $P \leq 0.05$.
**Significant at $P \leq 0.01$.

Table 5. The effect of soil acidity and organic fertilizers on crop productivity.

treatments where the mixture was grown in the soil applied with organic fertilizers. Fertilization with farmyard manure decreased the number and dry matter mass of weeds during the first 2 years after application. However, the use of green manure on the background of different rates of farmyard manure increased the number of weeds in the cereal crops of the rotation system. In the naturally acid soil, organic fertilization also reduced weed incidence in crops, especially in the lupine-oats mixture. The effect of organic fertilization was weaker in the winter wheat crop, where green manure was incorporated on the background of different rates of farmyard manure. Significantly higher number of weeds was established in these plots. However, reduction in the number of weeds was observed in the farmyard manure fertilization plots.

Fertilization with farmyard manure did not have any effect on the weed number in stands of spring barley. In all the experimental years, short-lived weeds predominated in the crops and accounted for 96.4% of the total weed number. Statistically significant correlations were determined between the agrochemical indicators of soil and the total number and mass of weeds in the rotation crops. Statistical analysis showed that in the first year of the crop rotation, with liming-induced reduction of soil acidity, the number of weed species declined ($r = -0.96**$). In all experimental years, the total number of weeds significantly depended also on the mobile aluminum content in the soil: $r = 0.92**$ in winter wheat crop, $r = 0.86**$ in lupine-oats mixture, and $r = 0.84**$ in spring barley crop.

Treatment	Winter wheat		Lupine-oats mixture		Spring barley	
	Number m^{-2}	Mass of DM g m^{-2}	Number m^{-2}	Mass of DM g m^{-2}	Number m^{-2}	Mass of DM g m^{-2}
Soil pH, liming (factor A)						
Unlimed soil (pH 4.1–4.3)	182.1	101.6	335.4	96.6	127.3	71.4
Limed soil (pH 5.8–6.0)	101.3**	32.0**	247.2**	57.9**	98.6	8.54*
Organic fertilizers (factor B)						
Without organic fertilizers	137.3	83.1	356.7	122.7	98.0	38.5
Green manure or plant residues	128.7	90.5	348.3	97.1	123.7	62.2
Farmyard manure 40 t ha^{-1}	92.7*	35.6**	256.7**	70.1*	112.0	28.9
Green manure (bkgd of FM 40)	177.7	42.1**	243.3**	45.3**	110.3	27.7
Farmyard manure 60 t ha^{-1}	116.3	41.4**	222.7**	63.7*	105.3	14.8*
Green manure (bkgd of FM 60)	197.7	108.0	320.2	64.5**	128.3	68.8
Interaction of factors A × B						
	ns	**	**	**	ns	**

ns, not significant; DM, dry matter.*Significant at $P \leq 0.05$.
**Significant at $P \leq 0.01$.

Table 6. The effect of soil acidity and organic fertilizers on the weed infestation in the crops of the rotation during the maturity stage.

The plowed layer (0–20 cm) of limed soil was significantly less contaminated with weed seeds compared with naturally acid soil.

The naturally acid soil without organic fertilization contained 7.5 times more weed seeds compared with limed soil. Irrespective of the soil pH, green manure increased soil contamination with weed seeds by 9.4%. However, fertilization with different rates of farmyard manure or incorporation of green manure on the background of different rates of farmyard manure resulted in 54.1–66.0% reduction in the weed seed bank in the soil. This might have been influenced by more active microbiological processes in the soil because of which part of weed seeds were more rapidly mineralized. In the total weed seed bank of not limed soils (pH 4.0–4.1), *Spergula arvensis* L. and or *Scleranthus annuus* L. accounted for 73.7%. In the total weed seed bank of limed soils (pH 6.4–6.8) nitrophilous weed *Chenopodium album* L accounted for 72.8%.

7. Conclusions

The findings of the long-term (more than half a century) liming and fertilizing experiments indicate that liming alone was less efficient for improvement of moraine loam soil Bathygleyic

Dystric Glossic Retisol acidity indicators than its combination with farmyard manure (FYM). The acidification of the soil was neutralized in the topsoil (Ah) and subsoil (ElB) up to 60 cm depth by a systematic soil liming with 1.0 rate every 7 years of powdered limestone in combination with the application of 60 t ha^{-1} of FYM every 3–4 years. The highest mobile P_2O_5 content 220 mg kg^{-1} was in the soil which had been limed and fertilized with 60 t ha^{-1} manure. Changes of nutrients caused by the long-term liming and manuring were established in the deeper horizons (ElB and ElBt) of soil profile as well. Liming is the most efficient practice used to reduce soil acidity, as it eliminates aluminum toxicity and increases calcium content. Under the effect of liming and FYM fertilization, a significant increase (1.9 and 1.2 times) occurred in the exchangeable Ca content in ElB and ElBt horizons, compared with the unlimed soil. Liming exerts an effect not only on exchangeable Ca accumulation in the soil profile but also promotes its leaching especially when combining liming, mineral and FYM fertilization. Liming and NPK fertilization as well as liming, and NPK and FYM fertilization of moraine loam resulted in slightly higher NO_3^- concentrations in the infiltration water during crop vegetation season compared with other practices. Soil pH determines the activity of biological processes occurring in the soil (CO_2 emission, atmospheric nitrogen fixation intensity, organic matter mineralization rate, distribution of beneficial microorganisms). The content of SOC was 1.45% in the unlimed treatment, while in periodically limed soil at 2.0 liming rate every 3–4 years, it was approximately 0.18 percentage points lower. The highest amount of soil organic carbon (1.91%) was obtained in the limed soil applied with FYM. Soil liming significantly decreased the amount of fulvic acids. Soil amendment with organic fertilizers significantly increased the amount of humic acids, which indicates an increase in organic matter fixation and its stability in the soil. The highest content of water stable macroaggregates (59.35%) was determined in the limed and manure-applied soil. Manure application alone did not have any effect on the formation of water stable aggregates in the topsoil. In limed soil, winter wheat grain yield was 1.8 times and barley grain yield 2.2 times higher than in acid soil. Soil acidity had a significant influence on crop weediness during the stage of maturity. In limed soil, the weed number decreased by 31.1% and their mass by 65.5%, compared to unlimed soil.

Acknowledgements

The chapter presents research findings, obtained through the long-term research program "Productivity and sustainability of agricultural and forest soils" implemented by Lithuanian Research Centre for Agriculture and Forestry. Also the chapter presents research findings, obtained through the project "The effect of long-term management of resources of varying intensity on the soils of different genesis and other components of agro-ecosystems (SIT-9/2015)" funded by the Research Council of Lithuania.

Author details

Danute Karcauskiene*, Regina Repsiene, Dalia Ambrazaitiene, Regina Skuodiene and Ieva Jokubauskaite

*Address all correspondence to: danute.karcauskiene@vezaiciai.lzi.lt

Vezaiciai Branch of Lithuanian Research Centre for Agriculture and Forestry, Vezaiciai, Lithuania

References

[1] Pengerud A, Stalnacke P, Bechmann M, Matniesen GB, Iital A, Koskiaho J, Kyllmac K, Lagzdins A, Povilaitis A. Temporal trends in phosphorus concentrations and losses from agricultural catchments in the Nordic and Baltic countries. Acta Agriculture Scandinavica, Section B—Soil and Plant Sciences. 2015;**65**:173-185. DOI: 10.1080/09064710.2014.993690

[2] Wang Y, Bolter M, Chang Q, Duttmann R, Schetz A, Petersen FJ, Wang Z. Driving factors of temporal variations in agricultural soil respiration. Acta Agriculture Scandinavica, Section B—Soil and Plant Sciences. 2015;**65**:589-604. DOI: 10.1080/09064710.2015.1036305

[3] Kinderiene I, Karcauskiene D. Effect of different crop rotations on soil erosion and nutrient losses under natural rainfall conditions in Western Lithuania. Acta Agriculture Scandinavica, Section B—Soil and Plant Sciences. 2012;**62**:199-205. DOI: 10.1080/09064710.2012.714400

[4] Jokubauskaite I, Slepetiene A, Karcauskiene D. Influence of different fertilization on the dissolved organic carbon, nitrogen and phosphorus accumulation in acid and limed soils. Eurasian Journal of Soil Sciences. 2015;**4**:76-143. DOI: 10.18393/ejss.91434

[5] Vigovskis J, Jermuss A, Svarta A, Sarkanbarde D. The changes of soil acidity in long-term fertilizer experiments. Zemdirbyste-Agriculture. 2016;**103**:129-134. DOI: 10.13080/z-a.2016.103.017

[6] Goulding KW. Soil acidification and the importance of liming agricultural soils with particular reference to United Kingdom. Soil Use Management. 2016;**32**:390-399. DOI: 10.111/sum.12270

[7] Eidukeviciene M, Volungevicius J, Marcinkonis S, Tripolskaja L, Karcauskiene D, Fullen MA, Booth CA. Interdisciplinary analysis of soil acidification hazard and its legacy effects in Lithuania. Natural Hazards and Earth System Sciences. 2010;**10**:1477-1485. DOI: 10.5194/nhess-10-1477-2010

[8] Ozeraitiene D. Possibility of optimal reaction and stable structure conservation in ecologically sensitive soils. Geoforma Ediciones. Man and Soil at the Third Millennium. 2002;**1**:727-735

[9] Devau N, Le Cadre E, Hinsinger P, Gerard F. A mechanistic model for understanding root-induced chemical changes controlling phosphorus availability. Annals of Botany. 2010;**105**:1183-1197. DOI: 10.1093/aob/mcg098

[10] Kunhikrishnan A, Thangarajan NS, Bolan NS, Xu Y, Mandal S, Gleeson DB, Seshadri B, Zaman M, Barton L, Tang C, Luo J, Dalal R, Ding W, Kirham MB, Naidu R. Functional relationships of soil acidification, liming and greenhouse gas flux. Advances in Agronomy. 2016;**139**:3-71. DOI: 10.1016/bs.agron.2016.05.001

[11] Wong MTF, Nortcliff S, Swift RS. Methods for determining the acid ameliorating capacity of plant residue compost, urban waste composts, farmyard manure and peat applied to tropical soils. Communication in Soil Science and Plant Analysis. 1998;**29**:2927-2937

[12] Yan F, Schubert S, Mengel K. Soil pH increase due to biological decarboxylation of organic acids. Soil Biology and Biochemistry. 1996;**28**:617-623

[13] Hue NV, Craddock GR, Adams R. Effects of organic acids on aluminium toxicity in sub-soils. Soil Science Society of America Journal. 1986;**50**:28-34

[14] Mokolobate MS, Haynes RJ. Comparative liming effect of four organic residues applied. Biology and Fertility of Soils. 2002;**35**:79-85. DOI: 10.1007/s00374-001-0439-z

[15] Teit R. Soil Organic Matter Biological and Ecological Effects. New-York: Jon Willey & Sons; 1990. 395 p

[16] Ossom EM, Rhykerd RL. Effects of lime on soil and tuber chemical properties and yield of sweetpotato [*Ipomoea batatas* (L.) lam.] culture in Swaziland. American—Eurasian Journal of Agronomy. 2008;**1**:1-5

[17] Mazvila J, Staugaitis G. Development of soil properties in Lithuania. In: Tripolskaja L, Masauskas V, Adomaitis T, Karcauskiene D, Vaisvila Z, editors. Management of Agroecosystem Components. Results of Long-Term Agrochemical Experiments. Lithuania: Akademija Press; 2010. pp. 31-47

[18] Pierce E, Warncke D. Soil and crop responce to variable rate liming to Michigan fields. Soil Science Society of America. 2000;**64**:774-780. DOI: 10.2136/sssaj2000.642774x

[19] Lalande R, Gagnon B, Royer I. Impact of natural and industrial liming materials on soil properties and microbial activity. Canadian Journal of Soil Science. 2009;**89**:209-222. DOI: 10.4141/CJSS08015

[20] Ge G, Li Z, Fan F, Chu G, Hou Z, Liang Y. Soil biological activity and their seasonal variations in response to long-term application of organic and inorganic fertilizers. Plant and soil. In: Zhu Y, editor. Vol. 326. Springer; 2010. p. 31

[21] Janusauskaite D, Kadziene G, Auskalniene O. The effect of tillage systems on soil microbiota in relation to soil structure. Polish Journal of Environmental Studies. 2013; **22**:1387-1139

[22] Redin M, Guenon R, Recous S, Schmatz R, Freitas LL, Aita C, Giacomini SJ. Carbon mineralization in soil of roots from twenty crop species, as affected by their chemical composition and botanical family. Plant and Soil. 2014;**378**:205-214. DOI: 10.1007/s11104-013-2021-5

[23] Kallenbach CM, Frey SD, Grandy AS. Direct evidence for microbial-derived soil organic matter formation and its ecophysiological control. Nature Communications. 2016;**7**:10. DOI: 10.1038/ncomms13630

[24] Rousk J, Philip I, Brookes C, Baath E. Contrasting soil pH effects on fungal and bacterial growth suggest functional redundancy in carbon mineralization. Applied and Environmental Microbiology. 2009;**75**:1589-1596

[25] Vaisvila Z, Tripolskaja L. The changes of soil biological activity due to the organic and mineral fertilizers impact. In: Tripolskaja L, Masauskas V, Adomaitis T, Karcauskiene D, Vaisvila Z, editors. Management of Agroecosystem Components. Results of Long-Term Agrochemical Experiments. Akademija Press; 2010. pp. 240-251

[26] Lei Z, Jian-ping G, Shi-ging W, Ze-Yang Z, Chao Z, Yongxiong Y, Can J. Mechanism of acid tolerance in a rhizobium strain isolated from *Peueraria lobata* (Willd.) Ohwi. Canadian Journal of Microbiology. 2011;**57**:514-524. DOI: 101139/w11-036

[27] Lapinskas E. Changes in Nitrogen and its Importance for Plants. Monograph, Akademija; 2008. 319 p

[28] Lapinskas E, Ambrazaitiene D, Piaulokaite-Motuziene L. Estimation of microbial properties in relation to soil acidity. Agronomijas Vestis, Latvian Journal of Agronomy. 2005;**8**:32-36

[29] Katterer T, Bolinder MA, Andren O, Kirchmann H, Menichetti L. Roots contribute more to refractory soil organic matter than above-ground crop residues, as revealed by a long-term field experiment. Agriculture, Ecosystems and Environment. 2011;**141**:184-192. DOI: 0.1016/j.agee.2011.02.029

[30] Poptrowska-Dlugosz A, Wilczewski E. Changes in enzyme activities as affected by green manure catch crops and mineral nitrogen fertilization. Zemdirbyste-Agriculture. 2014;**101**:139-146. DOI: 10.13080/z-a.2014.101.018

[31] Baumhardt RL, Stewart BA, Sainju UM. North American soil degradation: Processes, practices, and mitigating strategies. Sustainability. 2015;**7**:2936-2960. DOI: 10.3390/su7032936

[32] Lal R. Food security in a changing climate. Ecohydrology and Hydrobiology. 2013;**13**:8-21. DOI: 10.1016/j.ecohyd.2013.03.006

[33] Wiener WR, Bonan GB, Allison SD. Global soil carbon projections are improved by modelling microbial processes. Nature Climate Change. 2013;**3**:909-912. DOI: 10.1038/NCLIMATE1951

[34] Kirchmann H, Katterer T, Schon M, Borjesson G, Hamnér K. Properties of soils in the Swedish long-term fertility experiments: Changes in topsoil and upper subsoil at Örja and fors after 50 years of nitrogen fertilisation and manure application. Acta Agriculture Scandinavica, Section B—Soil and Plant Sciences. 2013;**63**:25-36. DOI: 10.1080/09064710.2012.711352

[35] Wen Y, Li H, Xiao J, Wang C, Shen Q, Ran W, He X, Zhou Q, Yu G. Insights into complexation of dissolved organic matter and Al (III) and nanominerals formation in soils under contrasting fertilizations using two-dimensional correlation spectroscopy and high resolution-transmission electron microscopy techniques. Chemosphere. 2014;**111**:441-449. DOI: 10.1016/j.chemosphere.2014.03.078

[36] Dagesse DF. Freezing cycle effects on water stability of soil aggregates. Canadian Journal of Soil Science. 2013;**93**:473-483. DOI: 10.4141/cjss2012-046

[37] Tang C, Rengel Z, Diatloff E, Gazey C. Responses of wheat and barley to liming on a sandy soil with subsoil acidity. Field Crops Research. 2003;**80**:235-244. DOI: 10.1016/S0378-4290(02)00192-2